Survival Analysis for Epidemiologic and Medical Research

A Practical Guide

This practical guide to the analysis of survival data written for readers with a minimal background in statistics explains why the analytic methods work and describes how to effectively analyze and interpret epidemiologic and medical survival data with the help of modern computer systems.

This text contains a variety of statistical methods that not only are key elements of survival analysis but also are central to statistical analysis in general. Techniques such as statistical tests, transformations, confidence intervals, and analytic modeling are discussed in the context of survival data but are, in fact, statistical tools that apply to many kinds of data. Similarly, discussions of such statistical concepts as bias, confounding, independence, and interaction are presented and also are basic to a broad range of applications. These topics make up essentially a second-year, one-semester biostatistics course in survival analysis concepts and techniques for nonstatisticians.

Steve Selvin is Professor of Biostatistics and Epidemiology at the University of California, Berkeley. He has taught on the Berkeley campus for 35 years and has authored or co-authored more than 200 scientific articles in the areas of applied statistics and epidemiology. He has received two university teaching awards and is a member of the ASPH/Pfizer Public Health Academy of Distinguished Teachers.

Practical Guides to Biostatistics and Epidemiology

Series advisors
Susan Ellenberg, *University of Pennsylvania School of Medicine*
Robert C. Elston, *Case Western Reserve University School of Medicine*
Brian Everitt, *Institute for Psychiatry, King's College London*
Frank Harrell, *Vanderbilt University Medical Centre*
Jos W. R. Twisk, *Vrije Universiteit Medical Centre, Amsterdam*

This is a series of short and practical but authoritative books for biomedical researchers, clinical investigators, public health researchers, epidemiologists, and nonacademic and consulting biostatisticians who work with data from biomedical, epidemiologic, and genetic studies. Some books are explorations of a modern statistical method and its application, others focus on a particular disease or condition and the statistical techniques most commonly used in studying it.

This series is for people who use statistics to answer specific research questions. The books explain the application of techniques, specifically the use of computational tools, and emphasize the interpretation of results, not the underlying mathematical and statistical theory.

Published in the series:
Applied Multilevel Analysis, by **Jos W. R. Twisk**
Secondary Data Sources for Public Health, by **Sarah Boslaugh**

Survival Analysis for Epidemiologic and Medical Research

A Practical Guide

Steve Selvin

University of California, Berkeley

CAMBRIDGE
UNIVERSITY PRESS

CAMBRIDGE
UNIVERSITY PRESS

University Printing House, Cambridge CB2 8BS, United Kingdom

One Liberty Plaza, 20th Floor, New York, NY 10006, USA

477 Williamstown Road, Port Melbourne, VIC 3207, Australia

314-321, 3rd Floor, Plot 3, Splendor Forum, Jasola District Centre, New Delhi - 110025, India

79 Anson Road, #06-04/06, Singapore 079906

Cambridge University Press is part of the University of Cambridge.

It furthers the University's mission by disseminating knowledge in the pursuit of
education, learning and research at the highest international levels of excellence.

www.cambridge.org
Information on this title: www.cambridge.org/9780521719377

First published 2008

A catalogue record for this publication is available from the British Library

Library of Congress Cataloging in Publication data
Selvin, S.
Survival analysis for epidemiologic and medical research : a practical guide / Steve Selvin
    p.   ;   cm. – (Practical guides to biostatistics and epidemiology)
Includes bibliographical references and index.
ISBN 978-0-521-89519-4 (hardback) – ISBN 978-0-521-71937-7 (pbk.)
1. Medicine – Research – Statistical methods.   2. Epidemiology – Research – Statistical
methods.   3. Survival analysis (Biometry)   I. Title.   II. Series.
[DNLM: 1. Survival Analysis.   2. Models, Statistical. WA 950 S469s 2008]
R853.S7S453   2008
610.72'7 – dc22   2007037910

ISBN 978-0-521-89519-4 Hardback
ISBN 978-0-521-71937-7 Paperback

For Liz and David

Contents

Overview

The description of survival analysis techniques can be mathematically complex. The primary goal of the following description, however, is a sophisticated introduction to survival analysis theory and practice using only elementary mathematics, with an emphasis on examples and intuitive explanations. The mathematical level is completely accessible with knowledge of high school algebra, a tiny bit of calculus, and a one-year course in basic statistical methods (for example, t-tests, chi-square analysis, correlation, and some experience with linear regression models). With this minimal background, the reader will be able to appreciate why the analytic methods work and, with the help of modern computer systems, to effectively analyze and interpret much of epidemiologic and medical survival data.

A secondary goal is the introduction (perhaps the review) of a variety of statistical methods that are key elements of survival analysis but are also central to statistical data analysis in general. Such techniques as statistical tests, transformations, confidence intervals, analytic modeling, and likelihood methods are presented in the context of survival data but, in fact, are statistical tools that apply to many kinds of data. Similarly, discussions of such statistical concepts as bias, confounding, independence, and interaction are presented in the context of survival analysis but also are basic to a broad range of applications.

To achieve these two goals, the presented material is divided into nine topics:

Chapter 5: Exponential survival time probability distribution
Chapter 6: Weibull survival time probability distribution
Chapter 7: Analysis of two-sample survival data
Chapter 8: General hazards model: parametric
Chapter 9: General hazards model: nonparametric

These topics make up essentially a second-year, one-semester biostatistics course. In fact, this course has been taught at the University of California, Berkeley as part of the biostatistics/epidemiology master of public heath degree major, at the Graduate Summer Institute of Epidemiology and Bio-statistics at Johns Hopkins Bloomberg School of Public Health, and at the Graduate Summer Session in Epidemiology at the University of Michigan.

All statistical methods are extensively illustrated with both analytic and graphical examples from the San Francisco Men's Health Study. This unique study was established in 1983 to conduct a population-based prospective investigation of the epidemiology and natural history of the newly emerging disease Acquired Immunodeficiency Syndrome (AIDS). The collected data are a source of valuable and comprehensive information about the AIDS epidemic in its earliest years. These data illustrate realistically the discussed statistical techniques. A "workbook" of noncomputer problems is included to further explore the practical side of survival analysis methods. Finally, a small amount of computer code gives a sense of survival analysis software. The statistical analysis system called "R" is chosen because it is extensive and fully documented and both the software and documentation can be obtained without cost (http://www.r-project.org).

Clearly many kinds of phenomena fail. Data collected to study the failure of equipment, machine components, numerous kinds of products, and the structural integrity of various materials are frequently analyzed with survival analysis techniques (sometimes called time-to-failure data and methods). For the following description of survival analysis, however, the terminology is by and large in terms of human mortality (survived/died). For example, rates are described in terms of mortality risk (risk of death). The language of human mortality is chosen strictly for simplicity. The theory and applications of the methods discussed are essentially the same regardless of the subject matter context. Using general terminology complicates explanations and is avoided to clearly focus on the statistical issues important in the analysis of epidemiologic and medical survival data.

It has been remarked (by Churchill Eisenhert) that the practical power of a statistical procedure is the statistical power multiplied by the probability that the procedure will be used. The material in this text has some of this same spirit. A number of analytic approaches are presented because they are simple rather than optimally efficient. For example, simple stratification procedures are suggested for estimation, exploring linearity of a variable, identifying the source of interactions, and assessing the proportionality of hazard functions. Also in the spirit of simplicity, all confidence intervals are set at the 95% level because other levels of significance are rarely used.

Steve Selvin, 2007

Rates and their properties

Rates are ratios constructed to compare the change in one quantity to the change in another. For example, postal rates are the price per unit weight for mailing a letter (price per ounce); miles divided by time produces a rate of speed (miles per hour). However, to understand and clearly interpret a rate applied to human survival data, a more detailed description is necessary. This description begins with Isaac Newton, who in the 17th century mathematically defined a rate and derived many of its properties.

The key to describing human survival, measured by rates of death or disease, is a specific function, traditionally denoted by $S(t)$, called the *survival function*. A survival function produces the probability of surviving beyond a specific point in time (denoted t). In symbols, a formal definition is

$$S(t) = P(\text{surviving from time} = 0 \text{ to time} = t)$$
$$= P(\text{surviving during interval} = [0, t])$$

or, equivalently,

$$S(t) = P(\text{surviving beyond time } t) = P(T \geq t).$$

Because $S(t)$ is a probability, it is always between zero and one for all values of t $(0 \leq S(t) \leq 1)$.

A simple survival function, $S(t) = e^{-0.04t}$, illustrates this concept (Figure 1.1). Perhaps such a function describes the pattern of 18th-century mortality for any age t (probability of living beyond age t). The probability of surviving beyond $t = 20$ years is, for example, $S(20) = P(T \geq 20) = e^{-0.04(20)} = 0.449$ (Figure 1.1). Similarly, this survival function dictates that half the population lives beyond 17.327 years. Thus,

$$P(\text{surviving beyond 17.327 years}) = S(T \geq 17.327) = e^{-0.04(17.327)} = 0.50.$$

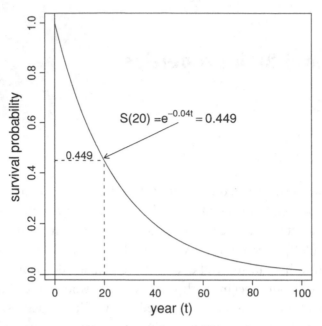

Figure 1.1.   A simple survival function—$S(t) = P(T \leq t) = e^{-0.04t}$.

To create a rate that does not depend on the length of the time interval, Newton defined an instantaneous rate as the change in $S(t)$ as the length of the time interval (denoted δ) becomes infinitesimally small. This version of a rate is *the derivative of the survival function $S(t)$ with respect to t* or, in symbols,

$$\text{the derivative of } S(t) = \frac{d}{dt} S(t).$$

The derivative of a function is a rich concept and a complex mathematical tool completely developed in a first-year calculus course. From a practical point of view, the derivative is closely related to the slope of a line between two points (an appendix at the end of the chapter contains a few details). That is, for two points in time (t and $t + \delta$), the derivative is approximately

$$\frac{d}{dt} S(t) \approx \frac{S(t + \delta) - S(t)}{\delta}$$

$$= \text{slope of a straight line between } S(t) \text{ and } S(t + \delta).$$

When the change in the survival function $S(t)[S(t)$ to $S(t + \delta)]$ is divided by the corresponding change in time t (t to $t + \delta$), one version of a rate becomes

$$\text{rate} = \frac{\text{change in } S(t)}{\text{change in time}} = \frac{S(t + \delta) - S(t)}{(t + \delta) - t} = \frac{S(t + \delta) - S(t)}{\delta}.$$

The proposed rate, constructed from two specific values of the survival function $S(t)$ and the length of the time interval δ, consists of the change (decrease) in the survival function $S(t)$ relative to the change (increase) in time (δ). For small values of δ, this rate (the slope of a line) hardly differs from an instantaneous rate. In the following, the slope of a line (one kind of rate) is frequently used to approximate the derivative of the survival function at a specific point in time, an instantaneous rate.

Newton's instantaneous rate is rarely used to describe mortality or disease data, because it does not reflect risk. A homicide rate, for example, of 10 deaths per month is easily interpreted in terms of risk only when it refers to a specific population size. A rate of 10 deaths per month in a community of 1,000 individuals indicates an entirely different risk than the same rate in a community of 100,000.

When the instantaneous rate $(d/dt)S(t)$ is divided by the survival function $S(t)$, it reflects risk. To measure risk, a relative rate is created, where

$$\text{instantaneous relative rate} = h(t) = -\frac{\frac{d}{dt}S(t)}{S(t)}.$$

Multiplying by -1 makes this relative rate a positive quantity, because $S(t)$ is a decreasing function (negative slope). An instantaneous relative rate $h(t)$ is usually called a *hazard rate* in human populations and a *failure rate* in other contexts. The same rate is sometimes called *the force of mortality* or *an instantaneous rate of death* or, from physics, *relative velocity*.

Two properties of a hazard rate complicate its application to collected data. The exact form of the survival function $S(t)$ must be known for all values of time t and the hazard rate is instantaneous. Knowledge is rarely available to unequivocally define $S(t)$ completely, instantaneous quantities are not intuitive, and interpretation frequently requires special mathematical/statistical tools.

Instead of an instantaneous rate, an average rate is typically used to measure risk, particularly from epidemiologic and medical survival data. Formally, a rate averaged over a time interval from t to $t + \delta$ is

$$\text{average rate} = \frac{S(t) - S(t + \delta)}{\int_t^{t+\delta} S(u)du}.$$

In more natural terms, an average rate over a specified time period is simply the proportion of individuals who died ("mean number of deaths") divided by the mean survival time for all individuals at risk during that period. Equally, an average rate is the total number of individuals who died divided by the total accumulated time at risk. Geometrically, the value in the numerator of an average rate is the decrease in the survival probability between the two points t and $t + \delta$. The value of the integral in the denominator is the area under the survival curve $S(t)$ between the same two points and equals the mean survival time of individuals who lived the entire interval or died during the interval.

For the survival function $S(t) = e^{-0.04t}$ and the time interval $t = 20$ to $t = 25$ years ($\delta = 5$ years), the proportion of individuals who died (mean number of deaths) is $S(20) - S(25) = e^{-0.80} - e^{-1.00} = 0.449 - 0.368 = 0.081$ (Figure 1.1). The mean survival time for all individuals at risk (area) during the interval 20 to 25 years ($\delta = 5$) is

$$\text{area} = \int_t^{t+\delta} S(u)du = \int_{20}^{25} e^{-0.04u}du$$

$$= \frac{e^{-0.04(20)} - e^{-0.04(25)}}{0.04} = \frac{0.449 - 0.368}{0.04} = \frac{0.081}{0.04}$$

$$= 2.036 \text{ person-years.}$$

Thus, the mean survival time lived by individuals who survived the entire five-year interval and those who died during the interval (20–25 years) is 2.036 years. A mean time at risk of 2.036 years makes the average mortality rate

$$\text{average rate} = \frac{\text{mean number of death}}{\text{mean survival time}} = \frac{e^{-0.80} - e^{-1.00}}{2.036} = \frac{0.081}{2.036}$$

$$= 0.040 \text{ deaths per person-year}$$

$$= 40 \text{ deaths per 1,000 person-years.}$$

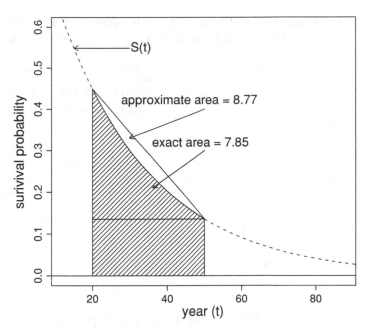

Figure 1.2. The geometry of an approximate average rate for the interval $t = 20$ to $t + \delta = 50$ (approximate rate $= 0.036$ and exact $=$ rate $= 0.040$).

In many situations, particularly in human populations, the area under the survival curve is directly and accurately approximated without defining the survival function $S(t)$, except at two points. When the survival function between the two points t and $t + \delta$ is a straight line, the area under the curve has a simple geometric form. It is a rectangle plus a triangle (Figure 1.2). Furthermore,

$$\text{area of the rectangle} = \text{width} \times \text{height} = ([t + \delta] - t) \times S(t + \delta)$$
$$= \delta S(t + \delta)$$

and

$$\text{area of the triangle} = \tfrac{1}{2} \text{ base} \times \text{altitude}$$
$$= \tfrac{1}{2}([t + \delta] - t) \times [S(t) - S(t + \delta)]$$
$$= \tfrac{1}{2}\delta[S(t) - S(t + \delta)],$$

Table 1.1. Approximate and exact areas for the time interval $t = 20$ and $t + \delta = 20 + \delta$ for the survival function $S(t) = e^{-0.04t}$ (exact rate $= 0.04$).

δ	t to $t + \delta$	$S(t)$	$S(t + \delta)$	$d(t)$	area*	area**	rate**
30	20 to 50.0	0.449	0.135	0.314	7.850	8.770	0.036
20	20 to 40.0	0.449	0.202	0.247	6.186	6.512	0.038
10	20 to 30.0	0.449	0.301	0.148	3.703	3.753	0.039
5	20 to 25.0	0.449	0.368	0.081	2.036	2.043	0.039
1	20 to 21.0	0.449	0.432	0.018	0.440	0.441	0.040
0.1	20 to 20.1	0.449	0.448	0.002	0.045	0.045	0.040

* $=$ exact $S(t)$
** $=$ approximate (straight line).

making the total area

area $=$ rectangle $+$ triangle
$$= \delta S(t + \delta) + \tfrac{1}{2}\delta[S(t) - S(t + \delta)] = \tfrac{1}{2}\delta[S(t) + S(t + \delta)].$$

Figure 1.1 displays the geometry for the survival function $S(t) = e^{-0.04t}$. For the interval $t = 20$ to $t + \delta = 25$ ($\delta = 5$), the area of the rectangle is $\delta S(25) = 5(0.368) = 1.839$ and the area of the triangle is $\tfrac{1}{2}\delta[S(20) - S(25)] = \tfrac{1}{2}(5)[0.449 - 0.368] = 0.204$, making the total area $1.839 + 0.204 = 2.043$ (mean time-at-risk during the interval). Again, the mean number of deaths is 0.0814. A measure of risk becomes the approximate average rate $= 0.0814/2.043 = 0.039$ (exact $= 0.04$) or 39 deaths per 1,000 person-years.

The approximate area is usually an accurate estimate of the exact area because the human survival curve in most situations is approximately a straight line over a specific and moderately small time interval. More simply, when a straight line and part of a survival function $S(t)$ are not very different, using an approximation based on a straight line works well [straight line $\approx$ $S(t)$]. Table 1.1 and Figure 1.2 illustrate this similarly for $t = 20$ years, where the exact average rate is 0.04 for all time intervals.

Because $S(t)$ represents the probability of surviving beyond time t, the difference $S(t) - S(t + \delta) = d(t)$ represents the probability of dying in the interval from t to $t + \delta$. In addition, the approximate area under the survival curve $S(t)$ has three equivalent forms, $\delta[S(t) - \tfrac{1}{2}d(t)]$ or $\delta[S(t + \delta) + \tfrac{1}{2}d(t)]$ or $\tfrac{1}{2}\delta[S(t) + S(t + \delta)]$, for the time interval t to $t + \delta$. All three expressions are the sum of the mean time lived by those who survived the entire interval [rectangle $= \delta S(t + \delta)$] plus the mean survival time lived by those who died [triangle $= \frac{1}{2}\delta d(t)$]. Therefore, to calculate the mean number of deaths and to approximate the mean time at risk, all that is needed is the values of $S(t)$ at the two points in time, namely t and $t + \delta$. The ratio of these two mean values is the average approximate mortality rate.

Example

Suppose that out of 200 individuals at risk, 100 individuals were alive January 1, 2004, and by January 1, 2006, suppose 80 of these individuals remained alive. In symbols, $t = 2004$, $t + \delta = 2006$ ($\delta = 2$ years), $S(2004) = 100/200 = 0.50$, and $S(2006) = 80/200 = 0.40$, making the proportion of the original 200 at-risk individuals who died during these two years $d(2004) = S(2004) - S(2006) = 0.50 - 0.40 = 0.10$ or $20/200 = 0.10$. The approximate area enclosed by the survival curve for this $\delta = 2$-year period is $\frac{1}{2} \cdot 2(0.50 + 0.40) = 0.90$ person-years (area). The average approximate rate becomes $R = (0.50 - 0.40)/0.90 = 0.10/0.90 = 0.111$ or, multiplying by 1,000, the rate is 111 deaths per 1,000 person-years. Rates are frequently multiplied by a large constant value to produce values greater than one (primarily to avoid small fractions). The mortality rate R reflects the approximate average risk of death over the period of time from 2004 to 2006 experienced by the originally observed 200 individuals. In addition, the total accumulated person-years lived by these 200 individuals during the two-year period is $200(0.90) = 180$ person-years because the mean years lived by these 200 individuals during the interval is 0.90 years. Therefore, the number who died $(100 - 80 = 200(0.10) = 20)$ divided by the total person-years (180) is the same approximate average rate,

$$\text{average rate} = R = \frac{\text{total deaths}}{\text{total person-years}} = \frac{20}{180} = \frac{0.10}{0.90} = 0.111.$$

The example illustrates the calculation of an approximate average rate free from the previous two constraints. It is not necessary to define the survival function $S(t)$ in detail and the rate is not instantaneous. The only requirements are that the two values $S(t)$ and $S(t + \delta)$ be known or accurately estimated and the survival curve be at least close to a straight line over the time period considered. Both conditions are frequently fulfilled by routinely collected human data providing a huge variety of mortality and disease rates (see the National Center for Health Statistics or the National Cancer Institute Web site—http://www.cdc.gov/nchs/ or http://www.nci.nih.gov).

It is important to note (or review) the equivalence of two ways to calculate a rate. An approximate average rate is calculated by dividing the mean number of deaths (the proportion of deaths) that occur during an interval by the mean survival time for that interval. That is, the ratio of means is

$$\text{approximate average rate} = \frac{\text{mean number of deaths}}{\text{mean survival time}}$$

$$= \frac{d(t)}{\frac{1}{2}\delta[S(t) + S(t + \delta)]}.$$

Or, more usually but less intuitively, the same rate calculated from a specific number of individuals (denoted l) in terms of deaths and total person-years is

$$\text{approximate average rate} = \frac{\text{total number of deaths}}{\text{total person-years at-risk}}$$

$$= \frac{l d(t)}{l\{\frac{1}{2}\delta[S(t) + S(t + \delta)]\}}.$$

These two rates are identical.

An approximate average rate is sometimes calculated by dividing the observed number of deaths by the number of individuals alive at the midpoint of the interval considered. For example, for the year 2000 in Marin County, California, there were 247,653 women alive halfway through the year and 494 deaths from cancer for the entire year. The annual average cancer mortality rate becomes 494 deaths divided by the midinterval count of 247,653 persons, and the approximate average rate $= (494/247,653) \times 100,000 = 199.5$ deaths per 100,000 person-years. This "short cut" is no more than an application of the fact that the midinterval population for l individuals is approximately the total accumulated person-years at risk or, in symbols, the midinterval population $l \times \delta S(t + \frac{1}{2}\delta)$ is approximately $l \times \frac{1}{2}\delta[S(t) + S(t + \delta)]$ and is exact when $S(t)$ is a straight line.

A number of ways exist to calculate an approximate average rate from mortality data based on the assumption that a straight line closely approximates the survival function. The following example illustrates three methods using

Table 1.2. U.S. mortality rates (all causes of death) age 65–74 for the years 1999, 2000, and 2001.

i	Year	Deaths (d_i)	Person-years ($pyrs_i$)	Rate/100,000
1	1999	387,437	16,167,771	2396.4
2	2000	376,986	16,100,428	2341.5
3	2001	367,128	15,969,452	2298.9
	Total	1,131,551	48,237,651	2345.8

U.S. mortality data for individuals aged 65 to 74 during the years 1999–2001 (Table 1.2).

Method 1:

$$\text{rate} = \frac{\sum d_i}{\sum pyrs_i} = \frac{1,131,551}{48,237,651} = 2,345.8 \text{ deaths per } 1000,000 \text{ person-years}$$

Method 2:

$$\text{rate} = \frac{d_2}{pyrs_2} = \frac{376,986}{16,100,428} = 2,341.5 \text{ deaths per } 100,000 \text{ person-years}$$

and

Method 3:

$$\text{rate} = \frac{\sum d_i}{3 \times pyrs_2} = \frac{1,131,551}{3 \times 16,1000,428}$$
$$= 2,342.7 \text{ deaths per } 100,000 \text{ person-years.}$$

The three methods produce essentially the same average mortality rate because the change in human mortality over short periods of time is usually close to linear.

Another frequent measure of risk is a probability. A probability, defined in its simplest terms, is the number of equally likely selected events (a subset) that might occur divided by the total number of all equally likely relevant events that could possibly occur (the entire set). In symbols, if $n[A]$ represents the number of selected events among a total of n equally likely events, then

$$\text{probability of event } A = P(A) = \frac{n[A]}{n}.$$

For example, the probability of death (denoted q) is $q = d/n$, where $n[A] = d$ represents the number of deaths among n individuals who could possibly have died. The complementary probability of surviving is $1 - q = p = (n - d)/n$. Notice the explicit requirement that all n individuals be members of a population with a proportion of q deaths and p survivors (next topic). Other, more rigorous definitions of probability exist, but this basic definition is sufficient for the following applications to survival analysis.

A probability is always zero (impossible event) or one (sure event) or between zero and one. In addition, a probability is unitless and does not depend directly on time. On the other hand, a rate can be any positive value, is not unitless (per person-time), and depends directly on time. Nevertheless, these two quantities are closely related. For an average approximate rate R and a probability q,

$$R = \frac{S(t) - S(t + \delta)}{\delta[S(t) - \frac{1}{2}d(t)]} = \frac{S(t)/S(t) - S(t + \delta)/S(t)}{\delta[S(t)/S(t) - \frac{1}{2}d(t)/S(t)]} = \frac{q}{\delta(1 - \frac{1}{2}q)}$$

and thus

$$q = \frac{\delta R}{1 + \frac{1}{2}\delta R},$$

where probability of death q is $d(t)/S(t)$ for the interval $(t, t + \delta)$. The probability of survival becomes $1 - q = p = S(t + \delta)/S(t)$. Note that q, and necessarily p, are conditional probabilities, conditional on being alive at time t. More specifically,

probability of death $= q = P(\text{death between } t \text{ and } t + \delta \mid \text{alive at time } t)$

$$= \frac{P(\text{death between } t \text{ and } t + \delta)}{P(\text{alive at time } t)} = \frac{d(t)}{S(t)}.$$

The probability of death or disease in human populations is almost always small ($p \approx 1$ or $q \approx 0$), making the relationship between a rate and a probability primarily a function of the length of the time interval δ. In symbols, the rate $= R \approx q/\delta$ when $\frac{1}{2}\delta q \approx 0$. When the period of time considered is one year, an average annual mortality rate and a probability of death typically produce almost identical values ($R \approx q$). These two quantities are more or less interchangeable and, particularly in the study of human mortality and disease, it often makes little practical difference which measure of risk is used.

For example, a ratio of rates and a ratio of probabilities hardly differ when applied to the same time interval. In symbols,

$$\text{rate ratio} = \frac{R_1}{R_0} \approx \frac{q_1/\delta}{q_0/\delta} = \frac{q_1}{q_0} = \text{relative risk.}$$

Under rather extreme conditions, a rate and a probability can differ considerably. For example, among 100 individuals, of whom 80 die in the first month during a disease outbreak and the remaining 20 survive the rest of the year ($\delta = 1$), the probability of death is $q = 80/100 = 0.8$ but the approximate average mortality rate is $R = 80/[20 + 0.5(1/12)(80)] = 0.80/[0.20 + 0.033] = 0.08/0.233 = 3.43$ deaths per person-year (area $= 0.233$). However, for the year considered, the probability of death is not small and the survival curve is not close to a straight line.

Statistical properties of the probability of death

When a rate is estimated from survival data, a fundamental assumption made about the sampled population is that the underlying probability of death (represented again by q) is at least approximately constant. "Constant," in this context, means that the probability q refers to a population made up of two outcomes (for example, died/survived or disease-present/disease-free) with a proportion of individuals q of one kind and a proportion of individuals $p = 1 - q$ of another kind. Under this condition, the properties of a sample of n individuals are described by a binomial probability distribution. Therefore, the probability that a sample of n independent individuals contains exactly d individuals who died and $n - d$ who survived is

$$P(D = d) = \binom{n}{d} q^d (1 - q)^{n-d} \qquad d = 0, 1, 2, \ldots, n$$

only when the probability of death q is constant.

These $n + 1$ probabilities, determined completely by the two parameters n and q, generate the properties of the binomially distributed variable represented by D. For example, the mean of the distribution of the count D is nq and its variance is given by the expression $nq(1 - q)$. The estimate of the binomial probability q is the number of sampled individuals who died divided by the total number sampled, denoted $\hat{q} = d/n$. The properties of this estimate also follow directly from the binomial probability distribution.

For example, the variance of the distribution of the estimate $\hat{q}$ is $q(1-q)/n$ and is naturally estimated by variance$(\hat{q}) = \hat{q}(1-\hat{q})/n$.

The variability of the distribution of the estimate $\hat{q}$, estimated by the expression $\hat{q}(1-\hat{q})/n$, reflects the sampling variation accompanying all statistical estimates. That is, another sample likely produces a different value of $\hat{q}$ because another sample will likely be made up of different individuals. It is this sample-to-sample variation that is measured by $q(1-q)/n$. It is this variation that is described by a binomial probability distribution. Occasionally the variation associated with the estimate $\hat{q}$ is erroneously attributed to the fact that individuals vary with respect to the probability of death. Variation of the probability of death among the sampled population members is an issue (to be discussed) but it is not the variation associated with a binomial distribution, which requires the probability represented by q to be constant. This distinction is important because the binomial distribution is central to the statistical description of probabilities and rates.

Two notable issues arise in applying a binomial distribution as part of describing a sample of survival data: the use of the normal distribution as an approximation and the consequences of assuming that the same constant probability q applies to all individuals within the sampled population when it does not.

Normal approximation

Statistical tests and confidence intervals based on a normal distribution are fundamental statistical tools used to assess the influence of sampling variation on an estimated value. In many situations, these tools apply to the estimated binomial probability $\hat{q}$. For example, an approximate 95% confidence interval is $\hat{q} \pm 1.960\sqrt{\text{variance}(\hat{q})}$ but requires the distribution of the estimate $\hat{q}$ to be at least approximately normal. This approximation works best when q is in the neighborhood of 0.5 and the sample size exceeds 30 or so ($n > 30$). These two requirements ensure that the distribution of the estimate $\hat{q}$ is close to symmetric and, therefore, is accurately approximated by a normal distribution. For survival data, particularly human survival data, the probability q typically refers to probabilities that are almost always small and in some cases extremely small. A consequence of a small probability is that the associated binomial probability distribution has a limited and positive range in the neighborhood of zero and is not symmetric. Because the normal distribution is symmetric and likely produces negative values near zero, it is no longer a directly useful approximation for a binomial distribution. Alternative approaches to evaluating an estimate $\hat{q}$ statistically when q is small employ exact methods or transformations.

Exact methods are conceptually complicated and numerically difficult but are available as part of statistical computer analysis systems. Transformations, on the other hand, require only a bit of calculation but, unlike exact methods, are conceptually simple. Transformations are created to make asymmetric distributions (sush as the binomial distribution with small q) approximately symmetric. For the transformed variable, the normal distribution once again becomes a useful approximation and normal-based tests and confidence intervals apply. In addition, these transformations are designed to always produce valid values for confidence interval bounds (for example, to never produce a negative bound for a probability).

Such a transformation of a small probability is the *logistic transformation*. A logistic transformation of an estimated probability $\hat{q}$ (denoted $\hat{l}$) is

$$\hat{l} = \log\left[\frac{\hat{q}}{1 - \hat{q}}\right] = \log\left[\frac{d}{n - d}\right].$$

The transformed estimate $\hat{l}$ has an unlimited range and a close to symmetric and, therefore, a more normal-like distribution. The value $\hat{l}$ is the logarithm of the odds, sometimes called the *log-odds* or *logit*. The odds are defined as the probability that an event occurs divided by the probability that the same event does not occur (the complementary event). The odds are a popular measure of risk used most often in gambling and epidemiology. To improve the accuracy (reduce the bias) of this logistic transformation, a value of one-half is added to the numerator and denominator, creating a less biased estimated log-odds,

$$\hat{l} = \log\left[\frac{d + \frac{1}{2}}{n - d + \frac{1}{2}}\right].$$

The estimated variance of the normal-like distribution of the estimate $\hat{l}$ is given by the expression

$$\text{variance}(\hat{l}) = \frac{(n + 1)(n + 2)}{n(d + 1)(n - d + 1)}.$$

The variance of the distribution of $\hat{l}$ is approximately variance$(\hat{l}) \approx 1/(d+1)$ when n is much larger than d, which is frequently the case for mortality and disease data (q is small). The origin of such a variance is discussed in Chapter 3.

The estimated probability of death from cancer among female residents of Marin County, California over the age of 30 is $\hat{q} = 494/247{,}900 = 0.001993$ or 199.3 per 100,000 at-risk women ($d = 494$ deaths among $n = 247{,}900$ women who were residents of Marin County at the beginning of the year 2000). Construction of a confidence interval from this estimate provides an example of applying a logistic transformation to mortality data (small q).

The estimated log-odds value is $\hat{l} = \log(494.5/247406.5) = -6.215$ with estimated variance of the distribution of $\hat{l}$ given by variance$(\hat{l}) = 0.00202$. The bounds of an approximate 95% confidence interval based on the estimated log-odds $\hat{l} = -6.215$ and the normal distribution, as usual, are

$$A = \text{lower bound} = \hat{l} - 1.960\sqrt{\text{variance}(\hat{l})}$$
$$= -6.215 - 1.960\sqrt{0.00202} = -6.303$$

and

$$B = \text{upper bound} = \hat{l} + 1.960\sqrt{\text{variance}(\hat{l})}$$
$$= -6.125 + 1.960\sqrt{0.00202} = -6.127.$$

A little algebra shows that $1/(l + e^{-l}) = \hat{q}$. Therefore, the log-odds 95% confidence interval bounds A and B calculated from the approximate normal distribution of $\hat{l}$ are identically transformed into the bounds associated with the estimated probability $\hat{q}$. The approximate 95% confidence interval bounds for the cancer rate in Marin County become

$$\text{lower bound} = \frac{1}{1 + e^{-A}} = \frac{1}{1 + e^{6.303}} = 0.001837$$

and

$$\text{upper bound} = \frac{1}{1 + e^{-B}} = \frac{1}{1 + e^{6.127}} = 0.002178$$

or (182.7, 217.8) per 100,000 at-risk women. As required, the probability $\hat{q}$ remains $1/(1 + e^{6.215}) = 0.001993$ or 199.3 deaths per 100,000 at-risk women. In addition, these log-odds calculated bounds will always be between

0 and 1. (A few details of the construction of confidence intervals based on transformed estimates are reviewed at the end of the chapter).

The logistic transformation similarly applies to the comparison of esti-mated probabilities from different populations (sometimes called the two-sample problem). For example, the probability of a cancer death in Marin County compared to the same probability for the rest of the state of California provides a formal evaluation of the observed excess risk expe-rienced in this county. The Marin County probability is again 199.3 cancer deaths per 100,000 at-risk women and the same probability for the rest of the state is 147.6 cancer deaths per 100,000 at-risk women (specifically, $[51,186/34,689,000] \times 100,000 = \hat{q} \times 100,000 = 147.6$). The correspond-ing logistic transformed estimates are $\hat{l}_{Marin} = -6.215$ and $\hat{l}_{state} = -6.517$. Applying the normal distribution approximation again provides an accurate assessment of the influence of sampling variation on the observed difference in log-odds transformed values. Specifically, the comparison takes the form

$$z = \frac{\hat{l}_{Marin} - \hat{l}_{state}}{\sqrt{variance(\hat{l}_{Marin} - \hat{l}_{state})}} = \frac{-6.215 - (-6.517)}{\sqrt{0.00202 + 0.00002}} = \frac{0.302}{0.045} = 6.680,$$

where the estimated $variance(\hat{l}_{Marin} - \hat{l}_{state}) = variance(\hat{l}_{Marin}) + vari-ance(\hat{l}_{state})$. For the comparison of cancer mortality risk between Marin County and the state as a whole, the estimated variance is $variance(\hat{l}_{Marin} - \hat{l}_{state}) = 0.00204$. The test statistic z has an approximately standard normal distribution when the underlying cancer mortality rates of Marin County and the state of California are the same, implying that the estimated log-odds values differ by chance alone. A significance probability (p-value) of $P(|Z| \geq 6.680|$ no difference$) < 0.001$ leaves little doubt that random variation is an unlikely explanation of the observed difference. The sig-nificance probability derived from the comparison of the more symmetric (normal-like) logit transformed probability (-6.215 compared to -6.517) equally applies to the comparison of the estimated probabilities themselves (199.3 compared to 147.6 deaths per 100,000). Both comparisons yield the identical p-value. In symbols, $P(|\hat{q}_{Marin} - \hat{q}_{state}| > 0|$ no difference$) = P(|Z| \geq 6.680|$ no difference$) < 0.001$.

This statistical test is consistent with the previous confidence interval con-structed from the Marin County cancer mortality data. The Marin County approximate 95% confidence interval (182.7, 217.8) defines a range of likely

Table 1.3. Four hypothetical groups ($n = 160$) heterogeneous for the probability q.

Group	n_i	d_i	q_i	v_i^*
Group 1	60	2	0.033	1.933
Group 2	50	4	0.080	3.680
Group 3	30	6	0.200	4.800
Group 4	20	8	0.400	4.800
Combined	160	20	0.125	17.500

* Variance of $d_i = v_i = n_i q_i (1 - q_i)$.

values for the underlying probability of a cancer death q (per 100,000 women) and does not include the estimated probability of death for the entire state (147.6/100,000). Thus, the statewide probability (or log-odds) is not a plausible value for Marin County from either perspective (test or confidence interval). The two approaches rarely give substantially different answers.

Homogeneity of the binomial probability q

Human populations are never perfectly homogeneous with respect to the probability of death or disease ($q = $ constant). Age-, race-, location-, and sex-specific samples of data are frequently collected, but the underlying probability of death remains heterogeneous to some extent even in these more highly stratified populations. The consequences of ignoring this residual heterogeneity are demonstrated by an example.

Suppose a population of 160 ($n = 160$) individuals consists of four groups heterogeneous for the probability q (defined in Table 1.3). A natural estimate of the probability q is $\hat{q} = d/n = 20/160 = 0.125$ ($d = \sum d_i$ and $n = \sum n_i$) combining the four groups. The estimated variance of the estimate $\hat{q}$ is $\hat{q}(1 - \hat{q})/n = (0.125)(0.875)/160 = 0.0007$. Both estimates completely ignore the heterogeneity of the probability q.

An estimate accounting for heterogeneity is the weighted average $\hat{q} = \sum n_i \hat{q}_i / \sum n_i = \sum d_i / n$ and again $\hat{q} = 0.125$. The estimated variance of $\hat{q}$ that accounts for the heterogeneity among the four groups is, however, reduced. The estimated variance is $\sum v_i / n^2 = 0.0006$. In symbols, the perhaps not very intvitive result emerges that

variance($\hat{q}$ | accounting for heterogeneity)

$\leq$ variance($\hat{q}$ | ignoring heterogeneity).

Not accounting for the heterogeneity of the probability q among a series of groups always produces a conservative estimate of the variability—conservative in the sense that the estimated variance is likely too large, causing statistical tests to have larger p-values or confidence intervals with greater lengths than would occur if the heterogeneity among groups were taken into account.

The difference in variability is entirely due to the differences among the $\hat{q}_i$-values. Specifically, the difference between the two estimated variances $\hat{q}(1 - \hat{q})/n$ and $\sum v_i/n^2$ is strictly a function of the differences among the $\hat{q}_i$-values. Or, in symbols, the expression for difference is $\sum n_i(\hat{q}_i - \hat{q})^2/n^2$. For the hypothetical example (Table 1.3), the heterogeneity of q among the four groups measured by $\sum n_i(\hat{q}_i - \hat{q})^2/n^2$ is 0.0001. The variance estimated by $\hat{q}(1 - \hat{q})/n$ is strictly correct only when $\hat{q}_i$ exactly equals $\hat{q}$ in all subgroups; otherwise it is biased upward. That is, the estimated variance of $\hat{q}$ is increased by ignoring heterogeneity because it is the sum of the estimated variance of $\hat{q}$ that accounts for heterogeneity and the variance of the values of $\hat{q}_i$ among the groups.

This artificial example is realistic in the sense that the bias arising from ignoring heterogeneity not only is conservative but is typically small. Therefore, not entirely accounting for heterogeneity in a sampled population, a reality in most applied situations, leads to statistical tests and estimated confidence intervals that are likely understated but not likely misleading. For example, the previous analysis of the Marin County cancer mortality data does not account for the heterogeneity of the probability of death within the county, producing a slightly biased confidence interval and statistical test.

Survival probabilities, average rates, and hazard rates: an example

Consider the survival experience of a population of individuals during the age interval 90 to 100 years. These individuals are envisioned as dying at random during the next 10 years. Thus, all individuals are equally likely to die at any time during the 10-year period. Postulating such a mortality pattern is not entirely unrealistic and has an important application to life table calculations (next chapter). In other situations, postulating a uniform risk of death among a subgroup of individuals is frequently part of describing more complicated survival patterns. Furthermore, even this simple illustration characterizes the fundamental relationships among a survival probability, an average mortality

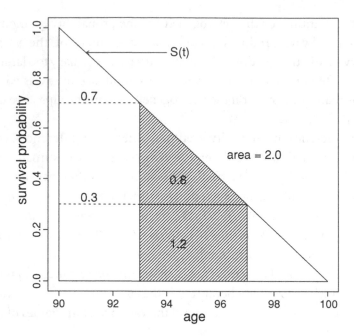

Figure 1.3.   Survival function $S(t)$ describing individuals dying uniformly during the age interval 90 to 100 years.

rate, and a hazard rate. More realistic situations differ in technical details but often differ little in principle.

When all individuals are equally likely to die at any time during the age interval from 90 to 100 years, then half the original individuals will live beyond age 95 and half will die before age 95. In fact, at any time (denoted t) during the 10 years, $[100 \times (100 - t)/10]$ percent will be alive and $[100 \times (t - 90)/10]$ percent will have died. For example, at age 97, 30% remain alive (70% have died). In symbols, the theoretical uniform survival probability function $S(t)$ is

$$S(t) = P(T \geq t) = \frac{b - t}{b - a}, \qquad \text{where } a \leq t \leq b.$$

In the present case, $a = 90$ and $b = 100$ years. Geometrically, the survival function (a continuous series of survival probabilities) is a straight line running from 1.0 at age 90 to 0.0 at age 100 with slope of $-1/(b - a) = -1/10 = -0.1$. As with all survival functions, the maximum value $S(a)$ is 1.0, occurring at age $= a = 90$, and the minimum value $S(b)$ is 0.0, occurring at age $= b = 100$ (Figure 1.3).

The average mortality rates of these 90-year-old individuals follows directly from the survival probabilities. Suppose l represents the total population of individuals at risk; then the total number of deaths between times t_1 and t_2 is

$$\text{number of deaths} = l \times [S(t_1) - S(t_2)] = \frac{l}{b-a}(t_2 - t_1)$$

and the total person-years these l individuals lived is

$$\text{person-years-at-risk} = l \times \text{area} = l \times \{\text{rectangle} + \text{triangle}\}$$
$$= l \times \left\{ (t_2 - t_1)S(t_2) + \tfrac{1}{2}(t_2 - t_1)[S(t_1) - S(t_2)] \right\}$$
$$= l \times \left\{ \frac{(t_2 - t_1)[b - \tfrac{1}{2}(t_1 + t_2)]}{b - a} \right\}.$$

For example, for $l = 1{,}000$ individuals at risk, the number of deaths between ages 93 and 97 would be

$$\text{number of deaths} = 1{,}000 \times [0.7 - 0.3] = \frac{1{,}000}{10} \times (97 - 93) = 400$$

and the total time lived between ages 93 and 97 would be

$$\text{person-years-at-risk} = 1{,}000 \times \frac{(97 - 93)\left[100 - \tfrac{1}{2}(93 + 97)\right]}{10} = 2{,}000.$$

The average mortality rate then becomes $R = 400/2{,}000 = 0.2$ (200 deaths per 1,000 person-years). Again, the total person-years lived between two points in time is the total number of individuals at risk multiplied by the mean survival time (the area under the survival curve or $l \times \text{area} = 1{,}000 \times 2.0 = 2{,}000$—Figure 1.3).

For these 90-year-olds, the expression for the average mortality rate for any interval t_1 to t_2 becomes

$$R = \text{average mortality rate} = \frac{\text{mean number of deaths}}{\text{mean person-years-at-risk}}$$
$$= \frac{\text{number of deaths}}{\text{total person-years-at-risk}}$$
$$= \frac{\frac{1}{b-a}(t_2 - t_1)}{l \times \left\{ \frac{(t_2 - t_1)[b - \tfrac{1}{2}(t_1 + t_2)]}{b - a} \right\}} = \frac{1}{b - \tfrac{1}{2}(t_1 + t_2)}.$$

For the example, the average mortality rate for individuals between the ages of 93 and 97 is again

$$R = \frac{1}{100 - \frac{1}{2}(93 + 97)} = 0.200 \text{ or } 200 \text{ deaths per } 1,000 \text{ person-years}$$

and the mortality rate for individuals between the ages of 97 and 100 years increases to

$$R = \frac{1}{100 - \frac{1}{2}(97 + 100)} = 0.667 \text{ or } 667 \text{ deaths per } 1,000 \text{ person-years.}$$

The average mortality rate indicates risk over a specific time interval. As the interval becomes smaller, the average mortality rate more accurately reflects the instantaneous hazard rate. When the length of the interval t_2 to t_1 ultimately becomes 0 or $t_1 = t_2 = t$ ($\delta = t_1 - t_2 = 0$), the two kinds of rates become identical. For the uniform mortality case when $\delta = 0$, the average rate becomes the hazard rate,

$$\text{hazard rate} = h(t) = \frac{1}{b - \frac{1}{2}(t + t)} = \frac{1}{b - t}.$$

This expression, derived directly from the definition of a hazard rate, is identical, or

$$h(t) = -\frac{\frac{d}{dt}S(t)}{S(t)} = \frac{\frac{1}{b - a}}{\frac{b - t}{b - a}} = \frac{1}{b - t}.$$

This hazard rate for the age interval 90 to 100 years is displayed in Figure 1.4.

The mean survival time for these 90-year-old individuals (entire population) is *mean years lived* $= \frac{1}{2}(b - a) = 0.5(100 - 90) = 5$ years. This mean value is geometrically the total area under the survival "curve" from age 90 to age 100 (area of the entire triangle—Figure 1.3). The total area under the survival curve is directly related to the total person-years of survival (denoted L). In symbols, $L = l \times$ mean years lived $= l \times \frac{1}{2}(b - a)$ person-years. In general,

$$\text{mean survival time} = \text{total area} = \frac{\text{total person-years}}{\text{the number of person-at-risk}} = \frac{L}{l}.$$

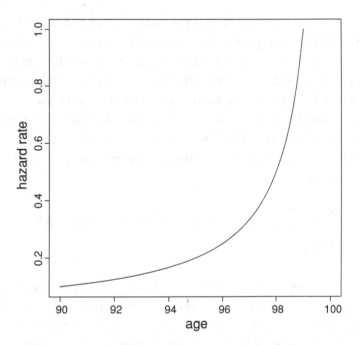

Figure 1.4. Hazard rate $h(t)$ for individuals dying uniformly during the age interval 90 to 100 years.

From this point of view, the mean survival time is calculated the same way as any mean value. It is the total amount of time lived (L) divided by the total number of individuals (l) who lived it.

The crude mortality rate (mortality rate for all individuals at risk over the entire interval from age 90 to 100 years) is

$$\text{crude rate} = \frac{\text{total number of deaths}}{\text{total person-years}} = \frac{d}{L} = \frac{l}{L}$$

$$= \frac{1}{\text{mean survival time}}.$$

Higher risk (rate) causes lower survival time (mean survival time) and vice versa. For the example, the mean survival time of five years makes the crude mortality rate $1/5 = 0.2$ deaths per person-year. Or, calculated directly, the crude mortality rate is

$$\text{crude mortality rate} = \frac{d}{L} = \frac{l}{L} = \frac{l}{l \times \frac{1}{2}(b - a)} = \frac{1}{5} = 0.2$$

for the entire interval. Notice that the total number of deaths (d) equals the total number of the original individuals at risk (l) when the entire time period is considered ($l = d$—everyone dies). A rate is not called crude because it is lacking or rudimentary. The statistical term "crude" applied to a rate simply means that it is not adjusted for influences from other factors.

The median survival time is that age at which half the original individuals are alive and half have died (denoted t_m). In symbols, when the survival probability $S(t_m) = 0.5$, the median survival time is t_m. For this uniform-mortality example,

$$S(t_m) = \tfrac{1}{2} = \frac{b - t_m}{b - a} \quad \text{and} \quad \text{median value} = t_m = b - \tfrac{1}{2}(b - a)$$
$$= \tfrac{1}{2}(b + a).$$

For the age period from 90 to 100 years, the median survival age is $t_m = \tfrac{1}{2}(100 + 90) = 95$ years, making the median years survived equal to 5. The mean and median survival times are equal because, for this special case of uniform risk, the pattern of mortality is symmetric (mean = median).

The probability of death during a specific age interval (again denoted q) is the number of individuals who died in the interval divided by the number who could have died (all at-risk individuals at the beginning of the interval). For the uniform mortality case, this conditional probability for the interval t_1 to t_2 is

$$\text{probability of death} = q = \frac{\text{total number of deaths}}{\text{total number of deaths that could occur}}$$

$$= \frac{l \times [S(t_1) - S(t_2)]}{l \times S(t_1)} = \frac{\dfrac{l}{b - a}(t_2 - t_1)}{\dfrac{l}{b - a}(b - t_1)} = \frac{t_2 - t_1}{b - t_1}.$$

For example, the probability of death during the age interval from 93 to 97 years is $q = 4/7 = 0.571$. More directly, when the original population consists of $l = 1,000$ individuals, among the 700 at-risk individuals aged 93, 400 died during the next four years; again $q = 400/700 = 0.571$. As before, the symbol q represents a conditional probability of death (conditional on being alive at age $t = 93$ or conditional on 700 at-risk individuals).

As with an average mortality rate and a survival probability, the probability of death (not surprisingly) is also related to the hazard rate. All three measures reflect the risk of death. Specifically, the hazard rate at any time t is

$$h(t) = \frac{1}{b - t},$$

and at time t_1

$$h(t_1) = \frac{1}{b - t_1} = \frac{1}{b - t_1} \times \frac{t_2 - t_1}{t_2 - t_1} = \frac{q}{t_2 - t_1},$$

producing another version of the previous rate/probability relationship ($R \approx q/\delta$). Specifically, at age 93, the hazard rate is $h(93) = 1/(100 - 93) = 1/7$ and the probability of death is $q = 4/7$; thus

$$h(93) = \frac{q}{t_2 - t_1} = \frac{4/7}{4} = \frac{1}{7}.$$

The relationship between a conditional probability of death and a hazard rate, as well as several other relationships illustrated, is useful in understanding more complex survival and hazard functions (future topics). Four relationships from this example that are important in other contexts are as follows:

1. An average rate approximately equals a hazard rate, particularly over a short interval of time, or

   average approximate rate $\approx$ hazard rate.

2. The hazard rate and the conditional probability of death are related, or

   $$\text{hazard rate} \approx \frac{\text{probability of death}}{\text{interval length}}.$$

3. The mean survival time is geometrically the area under the survival curve, or

   $$\text{mean survival time} = \frac{\text{total person-years}}{\text{total person-at-risk}}$$
   $$= \text{area under the survival curve.}$$

4. The approximate time lived by those who died (d) within the interval t_1 to t_2 is

   total person-time $\approx \frac{1}{2}(t_2 - t_1)d = \frac{1}{2}\delta d$ (a triangle).

APPENDIX I

Statistical tools: properties of confidence intervals

Consider the estimate of a generic parameter g, denoted $\hat{g}$. A normal-distribution-based approximate 95% confidence interval is

$$P[g - 1.960 S_{\hat{g}} \leq g \leq \hat{g} + 1.960 S_{\hat{g}}] = P[a \leq g \leq b] = 0.95,$$

where $S_{\hat{g}}$ represents the estimated standard deviation of the (at least approximate) normal distribution of the estimate $\hat{g}$.

A function applied to the lower (a) and upper (b) bounds produces an approximate 95% confidence interval for the same function of the parameter g, or, in symbols,

$$P[f(a) \leq f(g) \leq (b)] = 0.95.$$

For example, the estimate $\hat{g}$ produces

$P[\log(a) \leq \log(g) \leq \log(b)]$ as the 95% confidence interval for
    the logarithm of g,

$P[e^a \leq e^g \leq e^b]$ as the 95% confidence interval for the value of e^g, and

$P[a^2 \leq g^2 \leq b^2]$ as the 95% confidence interval for g^2.

The reverse is also true. A normal-based confidence interval constructed for a function of the parameter can be transformed to produce a confidence interval for the parameter itself. Specifically, if

$$P\left[f(\hat{g}) - 1.960 S_{f(\hat{g})} \leq f(g) \leq f(\hat{g}) + 1.960 S_{f(\hat{g})}\right] = P[A \leq f(g) \leq B]$$
$$= 0.95,$$

then an algebraic manipulation of the function $f(g)$ yields an approximate 95% confidence interval for the parameter g. For example, the estimate $f(\hat{g})$ produces

$$P[A \leq \log(g) \leq B] = P\left[e^A \leq e^{\log(g)} \leq e^B\right] = P[e^A \leq g \leq e^B]$$
$$= 0.95\,[f(g) = \log(g)],$$
$$P[A \leq g^2 \leq B] = P[\sqrt{A} \leq g \leq \sqrt{B}] = 0.95\,[f(g) = g^2], \text{ and}$$
$$P[A \leq \sqrt{\log(g)} \leq B] = P[A^2 \leq \log(g) \leq B^2] = P\left[e^{A^2} \leq g \leq e^{B^2}\right]$$
$$= 0.95\,[f(g) = \sqrt{\log(g)}].$$

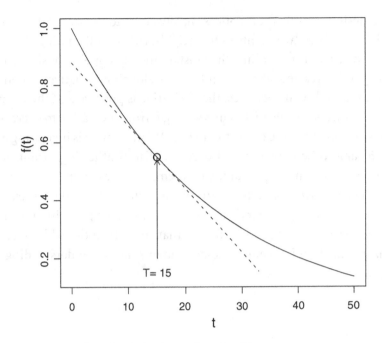

Figure 1.5. A representation of a derivative at the value T.

In short, a 95% confidence interval for the function $f(g)$ can be created from the 95% confidence interval for the parameter g and a 95% confidence interval for the parameter g can be created from the 95% confidence interval for $f(g)$.

APPENDIX II

Statistical tools: properties of a derivative of a function

Functions of continuous values can be represented by curves. Furthermore, the slope of a curve has no general definition and only one straight line touches a curve at a specific point. Combining these three properties produces the geometry of a derivative. It is the slope of the straight line at exactly the point where the line uniquely touches the curve. In Figure 1.5, this point is indicated within a circle and corresponds to the point $T = 15$.

The "slope" at this point is called the derivative of the function and can be viewed at the instantaneous "slope" at the point T. A natural property of a derivative is that it is approximately equal to the slope of a straight line calculated from two points in the neighborhood of the point T.

If the curve is increasing, the "instantaneous slope" is positive, and if the curve is decreasing, the "instantaneous slope" is negative. When a curve increases and then decreases, the derivative is positive and then is negative. At the exact point where the curve changes from increasing to decreasing (the maximum value), the derivative is zero. When a curve is increasing at exactly a 45° angle, the derivative is 1.0. A long list of other important properties exist and are a major part of the mathematics of calculus.

The relevance of a derivative to survival analysis is that a hazard function is the derivative of the survival function at a specific point in time divided by the value of the survival function at that point (multiplied by -1), and the hazard function is a central element analyzing and understanding survival data.

Life tables

The life table is perhaps the earliest statistical tool used to study human mortality rigorously. Early scientists Edmund Halley (1693) and John Graunt (1662) independently developed the first life tables from populations in Poland and England, respectively. A life table is essentially a highly organized description of age-specific mortality rates. Its importance has been reduced by modern methods (to be discussed), but it nevertheless remains fundamental to understanding survival data. A life table illustrates several basic statistical issues, particularly the roles of the two principal summaries of survival data, the survival function and the hazard rate.

Two kinds of life tables exist, called *cohort* and *current life tables*. Each kind has two styles, *abridged* and *complete*. A cohort life table is constructed from data accumulated by recording survival times from the birth of the first member of a population until the death of the last member. Collecting such cohort data is clearly impracticable in human populations. Cohort life tables primarily describe mortality patterns of small animal and insect populations. An abridged life table is based on a sequence of age intervals of any chosen length, typically five years. A current and complete life table is the subject of the following description. The life table components are derived from present-day observed mortality data (current) and applied to one-year age intervals (complete).

A life table describes the mortality experience of a cohort that does not exist. However, this theoretical cohort frequently provides valuable summaries of mortality patterns useful for comparing similarly constructed life table summaries from other groups or populations. For example, life table summaries derived from U.S. national mortality data collected in 1900 show that the mean lifetime for males was 46.6 years and for females was

48.7 years. Today, the same life table estimated summary values are 74.2 (males) and 79.3 (females) years.

Current, complete life table: construction

One-year, age-specific mortality rates produce a current, complete life table. Seven elements in this process are as follows:

Age interval (x to $x+1$): The symbol x represents the age of the individuals described by the life table. Each age interval is one year except the last, which is open-ended (for example, 90 years and older or 90^+ years).

Number alive (l_x): The symbol l_x represents the number of individuals alive (at-risk) at exactly age x. The number alive at age $x = 0$ (l_0) is set at some arbitrary number, such as 100,000, and occasionally called the *radix*.

Deaths (d_x): The symbol d_x represents the number of deaths between ages x and $x+1$ (one year).

Probability of death (q_x): The symbol q_x represents the conditional probability that a member of the life table cohort who is alive at age x dies before age $x+1$. In symbols,

$$P(\text{death before age } x + 1 \mid \text{alive at age } x) = q_x$$

and $q_x = d_x/l_x$. The complementary probability $p_x = 1 - q_x$ represents the conditional probability that an individual who is alive at age x survives beyond age $x+1$. It is necessary to distinguish clearly between the conditional survival probability p_x (conditional on a specific age) and the unconditional survival probability (denoted P_x). The symbol P_x represents the probability that an original member of the life table cohort survives beyond age x. In life table notation, these probabilities are $p_x = l_x/l_{x-1}$ (conditional) and $P_x = l_x/l_0$ (unconditional).

Years lived (L_x): The symbol L_x represents cumulative time lived by the entire cohort between the ages x and $x + 1$. Each individual alive at age x contributes to the total time lived during the next year; either one year if an individual lives the entire year or the proportion of the year lived when the individual dies within the one-year interval. The value of L_x

is the life table total person-years-at-risk accumulated by the l_x persons during the one-year age interval x to $x + 1$.

Total time lived (T_x): The symbol T_x represents the total time lived beyond age x by all individuals who are age x. The value T_x is the sum of the total person-years-at-risk lived in each age interval starting at age x. In symbols, $T_x = L_x + L_{x+1} + L_{x+2} + \cdots$. The accumulated time lived, T_x, is primarily a computational step in the life table construction.

Expectation of life (e_x): The symbol e_x represents the mean number of additional years lived by those members of the life table cohort who are age x. Computationally, $e_x = T_x/l_x$.

The total time lived by l_x persons in a specific one-year age interval (person-years-at-risk) is given by the expression

$$L_x = (l_x - d_x) + \bar{a}_x d_x.$$

The quantity $(l_x - d_x)$ represents the number of individuals who survived the entire interval (from x to $x + 1$). Another symbol for the same quantity is l_{x+1}. These individuals contribute $(l_x - d_x)$ person-years to the total L_x (one year for each person). Individuals who died within the interval each contribute the part of the year they were alive to the total survival time. Typically the exact times of death for these individuals are not available, particularly for age-specific mortality data extracted from public databases. However, for all but two age intervals, the mean survival time contributed by those who died (denoted $\bar{a}_x$) is essentially one-half year ($\bar{a}_x = 0.5$). The value 0.5 years is a dependable estimate because, for nearly human populations, individuals who are not extremely young die randomly or close to randomly throughout a single year. In general, no reason exists for an individual to die at a particular time during his or her next year of life. The difference between the probability of dying at the beginning and the end of a one-year age interval is small and unimportant for nearly all ages. The person-years contributed to the total survival time by those persons who died are, therefore, accurately estimated by $\bar{a}_x d_x = 0.5 d_x$ person-years (one-half year, on average, for each death—Chapter 1).

Two mean values of $\bar{a}_x$ fail to be even close to 0.5. They are the first ($\bar{a}_0$) and the last ($\bar{a}_{x+}$) values. The first value, $\bar{a}_0$, is usually set at 0.1. During the first year of life, most deaths occur within a few days after birth. That is, infant deaths do not occur randomly throughout the first year of life. Empirically, the mean time lived by those infants who died is about one month or $\bar{a}_0 = 0.1$ years. Therefore, the person-years contributed to the first-year total survival time L_0 by the d_0 infants who died before they were one year old are estimated by $0.1d_0$ person-years (0.1 years per death).

Mortality rates of extremely old individuals are rarely recorded in detail. After age 80 or so, ages are reported so unreliably that the associated age-specific mortality rates become equally unreliable. For this reason and the dramatically decreasing number of individuals at risk, the last several age intervals of a life table are combined into a single open-ended interval. For the complete U.S. life table describing the male mortality pattern for the year 2000 (Table 2.1), the last age interval is not one year but consists of all individuals who lived beyond age 90 (denoted 90^+). The mean time lived beyond age 90 by all individuals who reached the age of 90 (denoted $\bar{a}_{90^+}$) must be estimated directly from the age-specific rates because the last life table age interval does not have a specific end point.

For the last age interval (denoted x^+), the life table mortality rate (denoted r_{x^+}) is

$$\text{life table mortality rate (interval } x^+) = r_{x^+} = \frac{d_{x^+}}{L_{x^+}} = \frac{l_{x^+}}{L_{x^+}}$$

because all individuals alive at age x^+ ultimately die ($l_{x^+} = d_{x^+}$). The calculation of the total person-years l_{x^+} requires special attention. Substituting the observed current mortality rate (R_{x^+}) for the life table mortality rate (r_{x^+}) yields

$$R_{x^+} = \frac{l_{x^+}}{L_{x^+}} \quad \text{or} \quad L_{x^+} = \frac{l_{x^+}}{R_{x^+}}.$$

The value l_{x^+} represents the total person-years lived beyond age x^+ by the l_{x^+} individuals who reached age x^+. Then, the mean survival time for the last interval becomes

$$\bar{a}_{x^+} = \frac{L_{x^+}}{l_{x^+}} = \frac{1}{R_{x^+}}$$

and, to repeat, R_{x^+} represents the direct observed current mortality rate for individuals at risk during the last open-ended age interval. For example, if $R_{90^+} = 0.2$, then $\bar{a}_{90^+} = 1/R_{90^+} = 1/0.2 = 5.0$ years. Once again, the reciprocal of a rate equals the mean survival time (Chapter 1). Thus, the mean survival time $\bar{a}_{x+}$ and the total person-years-at-risk $L_{x+} = \bar{a}_x l_{x+}$ are estimated from the current mortality rate for the last age interval.

To summarize: for each age interval, the total person-years lived by those individuals at risk are

$$L_0 = (l_0 - d_0) + 0.1 d_0 \quad \text{for the } l_0 \text{ individuals of age 0 (newborn infants),}$$
$$L_x = (l_x - d_x) + 0.5 d_x \quad \text{for the } l_x \text{ individuals of ages } 1, 2, \ldots, (x^+ - 1),$$

and

$$L_{x+} = \frac{l_{x+}}{R_{x+}} = \bar{a}_{x+} l_{x+} \quad \text{for the } l_{x+} \text{ individuals of age } x^+$$
$$\text{(last and open-ended interval).}$$

Current mortality rates R_x also provide estimates of the probability of death for each age interval, namely estimates of each conditional probability q_x. Again, substituting the current age-specific mortality rate (R_x) for the life table mortality rate (r_x), the life table conditional probabilities of death q_x are estimated from current data. In symbols,

$$r_x = \frac{d_x}{L_x} = \frac{d_x}{l_x - \bar{a}_x d_x} = \frac{q_x}{1 - \bar{a}_x q_x} = R_x \quad \text{because} \quad q_x = \frac{d_x}{l_x}.$$

Solving for q_x gives the estimated conditional probability of death in the age interval x to $x + 1$ as

$$q_x = \frac{R_x}{1 + \bar{a}_x R_x},$$

based on the current age-specific mortality rate R_x. For example, the observed U.S. male mortality rate in the year 2000 (Table 2.1) for the first year of life is $R_0 = 15{,}612/1{,}949{,}017 = 0.00801$, making the probability of death $q_0 = 0.008$. As noted earlier, the average human mortality rate and the probability of death are essentially the same for a one-year interval $(R \approx q/\delta = q)$.

The estimation of q_0 is the first step in life table construction. The number of life table deaths occurring during the first year becomes $d_0 = l_0 q_0 = 100{,}000(0.00801) = 801$ when l_0 is set at the arbitrarily selected value of 100,000 persons-at-risk. Thus, the number of life table survivors who enter the second year is $l_1 = 100{,}000 - 801 = 99{,}199$. The total person-years lived by the 100,000 individuals (l_0) during the first year becomes $L_0 = (100{,}000 - 801) + 0.1(801) = 99{,}279$. The same kind of calculation yields values $q_1, d_1,$

Table 2.1. Life table: U.S. white males.

Interval	Population	Deaths	q_i	d_i	l_i	L_i	T_i	e_i
0–1	1,949,017	15,612	0.0080	801	100,000	99,279	7,420,473	74.2
1–2	1,953,105	1090	0.0006	55	99,199	99,168	7,321,194	73.8
2–3	1,938,990	746	0.0004	38	99,144	99,123	7,222,026	72.8
3–4	1,958,963	558	0.0003	28	99,106	99,092	7,122,903	71.9
4–5	2,010,658	430	0.0002	21	99,078	99,067	7,023,811	70.9
5–6	2,031,072	430	0.0002	21	99,056	99,046	6,924,744	69.9
6–7	2,058,217	380	0.0002	18	99,035	99,026	6,825,698	68.9
7–8	2,109,868	347	0.0002	16	99,017	99,009	6,726,672	67.9
8–9	2,137,829	349	0.0002	16	99,001	98,993	6,627,663	66.9
9–10	2,186,291	344	0.0002	16	98,985	98,977	6,528,670	65.0
10–11	2,191,244	390	0.0002	18	98,969	98,960	6,429,693	64.0
11–12	2,108,157	441	0.0002	21	98,952	98,941	6,330,733	63.0
12–13	2,087,228	462	0.0002	22	98,931	98,920	6,231,791	63.0
13–14	2,054,008	533	0.0003	26	98,909	98,896	6,132,871	62.0
14–15	2,078,560	725	0.0003	34	98,883	98,866	6,033,975	61.0
15–16	2,065,127	1019	0.0005	49	98,849	98,824	5,935,109	60.0
16–17	2,048,582	1493	0.0007	72	98,800	98,764	5,836,285	59.0
17–18	2,091,280	1952	0.0009	92	98,728	98,682	5,737,521	58.1
18–19	2,087,853	2440	0.0012	115	98,636	98,578	5,638,839	57.1
19–20	2,107,162	2793	0.0013	131	98,521	98,456	5,540,260	56.2
20–21	2,071,220	2812	0.0014	133	98,390	98,324	5,441,805	55.3
21–22	1,965,673	2855	0.0015	143	98,257	98,185	5,343,481	54.4
22–23	1,921,549	2649	0.0014	135	98,114	98,047	5,245,296	53.5
23–24	1,875,400	2546	0.0014	133	97,979	97,913	5,147,249	52.5
24–25	1,853,972	2512	0.0014	132	97,846	97,780	5,049,336	51.6
25–26	1,905,899	2405	0.0013	123	97,714	97,652	4,951,557	50.7
26–27	1,832,383	2349	0.0013	125	97,590	97,528	4,853,905	49.7
27–28	1,914,947	2483	0.0013	126	97,465	97,402	4,756,377	48.8
28–29	2,010,807	2561	0.0013	124	97,339	97,277	4,658,975	47.9
29–30	2,134,724	2821	0.0013	128	97,215	97,151	4,561,697	46.0
30–31	2,174,238	2765	0.0013	123	97,087	97,025	4,464,546	45.0
31–32	2,019,782	2889	0.0014	139	96,963	96,894	4,367,521	45.0
32–33	2,008,877	2918	0.0015	141	96,825	96,755	4,270,627	44.1
33–34	2,018,017	3166	0.0016	152	96,684	96,608	4,173,873	43.8
34–35	2,100,855	3533	0.0017	162	96,533	96,452	4,077,264	42.2
35–36	2,265,621	3861	0.0017	164	96,370	96,288	3,980,813	41.3
36–37	2,247,529	4247	0.0019	182	96,206	96,116	3,884,524	40.4

Interval	Population	Deaths	q_i	d_i	l_i	L_i	T_i	e_i
37–38	2,250,122	4624	0.0021	197	96,025	95,926	3,788,409	39.4
38–39	2,268,083	5099	0.0022	215	95,828	95,720	3,692,482	38.5
39–40	2,287,341	5393	0.0024	225	95,612	95,500	3,596,762	37.6
40–41	2,352,606	6003	0.0025	243	95,387	95,266	3,501,262	36.7
41–42	2,213,034	6292	0.0028	270	95,144	95,009	3,405,997	35.8
42–43	2,256,543	6731	0.0030	283	94,874	94,733	3,310,988	34.9
43–44	2,178,451	7339	0.0034	318	94,592	94,432	3,216,255	34.0
44–45	2,128,468	7680	0.0036	340	94,273	94,104	3,121,822	33.1
45–46	2,151,115	8170	0.0038	356	93,934	93,756	3,027,719	32.2
46–47	2,009,570	8579	0.0043	399	93,578	93,378	2,933,963	31.3
47–48	1,976,128	9016	0.0046	424	93,179	92,967	2,840,584	30.5
48–49	1,909,672	9531	0.0050	462	92,755	92,524	2,747,617	29.6
49–50	1,843,021	9825	0.0053	491	92,293	92,048	2,655,093	28.8
50–51	1,871,638	10,256	0.0055	502	91,802	91,552	2,563,046	27.9
51–52	1,769,463	10,614	0.0060	546	91,301	91,028	2,471,494	27.1
52–53	1,815,785	11,488	0.0063	572	90,755	90,469	2,380,466	26.2
53–54	1,778,423	12,364	0.0069	625	90,182	89,870	2,289,998	25.5
54–55	1,372,415	10,555	0.0077	686	89,558	89,215	2,200,128	24.6
55–56	1,386,859	11,498	0.0083	734	88,871	88,505	2,110,913	23.7
56–57	1,375,187	12,334	0.0089	787	88,138	87,744	2,022,408	22.0
57–58	1,384,196	13,597	0.0098	854	87,351	86,924	1,934,664	22.1
58–59	1,222,709	13,356	0.0109	940	86,497	86,027	1,847,740	21.4
59–60	1,139,778	13,640	0.0119	1018	85,557	85,048	1,761,713	20.6
60–61	1,111,560	14,152	0.0127	1070	84,539	84,005	1,676,665	19.8
61–62	1,061,679	14,657	0.0137	1144	83,470	82,898	1,592,661	19.1
62–63	1,033,865	15,832	0.0152	1251	82,325	81,700	1,509,763	18.3
63–64	971,203	16,511	0.0169	1367	81,074	80,391	1,428,063	17.6
64–65	958,320	17,744	0.0183	1462	79,708	78,976	1,347,672	16.9
65–66	950,651	18,614	0.0194	1517	78,245	77,487	1,268,696	16.2
66–67	864,156	18,960	0.0217	1665	76,728	75,896	1,191,209	15.5
67–68	874,079	20,658	0.0234	1753	75,063	74,186	1,115,313	14.7
68–69	856,145	22,082	0.0255	1867	73,310	72,376	1,041,127	14.2
69–70	855,331	23,621	0.0272	1946	71,443	70,470	968,751	13.6
70–71	844,517	25,690	0.0300	2082	69,497	68,456	898,281	12.9
71–72	798,517	27,007	0.0333	2242	67,414	66,293	829,826	12.3
72–73	791,164	29,167	0.0362	2359	65,172	63,993	763,532	11.7
73–74	751,433	30,297	0.0395	2483	62,813	61,572	699,540	11.1

(cont.)

Table 2.1 (cont.)

Interval	Population	Deaths	q_i	d_i	l_i	L_i	T_i	e_i
74–75	717,281	31,312	0.0427	2577	60,331	59,042	637,968	10.6
75–76	695,865	32,971	0.0463	2673	57,753	56,417	578,926	10.1
76–77	647,773	33,727	0.0507	2795	55,080	53,683	522,509	9.5
77–78	599,742	34,530	0.0560	2926	52,285	50,822	468,827	8.0
78–79	579,368	35,746	0.0599	2954	49,359	47,882	418,005	8.5
79–80	512,708	36,353	0.0685	3178	46,405	44,816	370,123	7.0
80–81	467,013	35,604	0.0734	3175	43,227	41,640	325,307	7.5
81–82	406,546	33,516	0.0792	3171	40,053	38,467	283,667	7.1
82–83	364,815	34,053	0.0892	3289	36,881	35,237	245,200	6.6
83–84	317,289	32,350	0.0970	3259	33,592	31,963	209,964	6.3
84–85	279,234	31,369	0.1064	3226	30,333	28,720	178,001	5.9
85–86	244,874	30,401	0.1169	3169	27,107	25,523	149,281	5.5
86–87	204,981	28,512	0.1301	3113	23,938	22,382	123,758	5.2
87–88	173,520	26,109	0.1399	2914	20,825	19,368	101,376	4.9
88–89	139,395	23,161	0.1534	2748	17,911	16,537	82,008	4.6
89–90	113,731	20,694	0.1668	2529	15,163	13,899	65,471	4.3
90+	350,497	85,865	1.0000	12,634	12,634	51,572	51,572	4.1

and L_1 for the second year of life, based on the observed age-specific mortality rate R_1 and the life table value l_1 for the second year of life. The current U.S. male mortality rate for the second year of life, $R_1 = 1090/1,953,105 = 0.0006$, translates into the conditional probability of death $q_1 = 0.0006$. Thus, the number of life table deaths during the second year is $d_1 = l_1 q_1 = 99,199(0.0006) = 55$ because $l_1 = 99,199$ individuals are at risk after the first year. The number of life table survivors is $l_2 = 99,199 - 55 = 99,144$ and the total person-years lived by the 99,199 individuals at risk from year one to year two becomes $L_1 = (99,199 - 55) - 0.5(55) = 99,168$. This same calculation is sequentially repeated for each one-year age interval up to age 89 (Table 2.1).

For the last and open-ended interval (age 90$^+$), the U.S. male mortality rate is $R_{90+} = 85,856/350,497 = 0.245$ and $l_{90+} = l_{89} - d_{89} = 15,163 - 2,529 = 12,634$ persons-at-risk. All 12,634 of these 90-year-old individuals will die during the final interval ($l_{90+} = d_{90+} = 12,634$ or $q_{90+} = 1.0$). Furthermore, the mean additional years of lifetime lived by these individuals is $\bar{a}_{90+} = 1/0.245 = 4.082$ years, making $L_{90}{}^+ = 12{,}634(4.082) = 12{,}634/0.245 = 51{,}572$ person-years lived during the final open-ended age interval 90^+ by the 12,634 individuals who lived beyond age 90. Thus, a current life table describes the mortality experience of a hypothetical cohort of l_0 individuals from "birth to death" as if they had exactly the same U.S. mortality rates for the next 100 years or so that were used to construct the life table (R_x = observed current cross-sectional mortality rates become r_x = theoretical life table longitudinal cohort mortality rates, or $R_x = r_x$).

Two remaining life table summaries (T_x and e_x) are constructed from the L_x and l_x values. The total lifetime lived beyond age x is $T_x = \sum L_i$ for $i = x, x+1, \ldots, x^+$. For the example (Table 2.1), total person-years lived by the $l_0 = 100{,}000$ members of the life table cohort from birth to death is $T_0 = 99{,}279 + 99{,}168 + \cdots + 51{,}572 = 7{,}420{,}473$ person-years ($x = 0$ to 90^+ years) and from age 60 years to death, $T_{60} = 84{,}005 + 82{,}898 + \cdots + 51{,}572 = 1{,}676{,}665$ person-years ($x = 60$ to 90^+ years).

The e_x-value summarizes life table survival in terms of mean additional years of remaining lifetime from age x. This mean value is calculated just like any mean value. It is

$$e_x = \text{mean additional years} = \frac{\text{total person-years lived beyond age } x}{\text{number of persons of age } x} = \frac{T_x}{l_x}.$$

The most fundamental single summary of life table mortality experience is the mean years of life of a newborn infant (denoted e_0). For the 2000 U.S. life table cohort, the mean years lived by a newborn male is $e_0 = T_0/l_0 = 7{,}420{,}473/100{,}000 = 74.205$ years (Table 2.1). For a U.S. male of age 60 (Table 2.1, again), the mean years of lifetime remaining is calculated in the same way. Specifically, the mean value is $e_{60} = T_{60}/l_{60} = 1{,}676{,}665/84{,}539 = 19.8$ years. Or, the typical life table 60-year-old male lives to age 79.8 years. Table 2.2 displays the mean years of additional life (e_0) of a newborn male child for seven selected countries based on current life tables.

The life table crude mortality rate and the mean years lived by a newborn e_0 are related by

$$\text{crude mortality rate} = \frac{\text{total deaths}}{\text{total person–years}} = \frac{d_0}{T_0} = \frac{l_0}{T_0} = \frac{1}{e_0}.$$

Table 2.2. Selected countries
and the life table mean survival
times for newborn males.

Country	e_0
China	66.9
Congo	48.5
India	63.3
Japan	77.6
Mexico	63.8
Norway	76.0
United States	74.2

Source: National Center for Health
Statistics, year 2000.

Specifically, for the year 2000 life table cohort,

$$\text{crude mortality rate} = \frac{100,000}{7,420,473} = \frac{1}{74.205} = 0.135$$

or 1350 deaths per 100,000 person-years (Table 2.1). For 60-year-old individuals, the life table mean lifetime remaining is 19.8 years and, therefore, the life table crude mortality rate is $1/19.8 = 0.0505$ or 5050 deaths per 100,000 person-years. The risk of death (crude mortality rate) is the reciprocal of the mean lifetime (e_x). As expected, higher risk reduces survival time.

The life table survival function from Table 2.1 (actually a series of 92 connected survival probabilities) is displayed in Figure 2.1. Life table survival probabilities directly calculated from the l_x values are:

life table probability of surviving beyond age $x = P_x$

$$= \prod(1 - q_i) = \prod p_i = \frac{l_1}{l_0} \times \frac{l_2}{l_1} \times \frac{l_3}{l_2} \times \cdots \times \frac{l_x}{l_{x-1}} = \frac{l_x}{l_0}$$

because the conditional and interval-specific survival probability is $p_i = l_i/l_{i-1}$. The unconditional survival probability P_x is the product of a series of conditional survival probabilities p_i and is the life table probability of surviving from birth to age x, or, more simply, the number of persons of age x divided by the total number of person at risk ($n[x] = l_x$ divided by $n = l_0$—Chapter 1). The survival curve for U.S. white males (Figure 2.1) displays the typical pattern for human mortality across the entire life span. A small dip during the first year of life due to high infant mortality is followed by

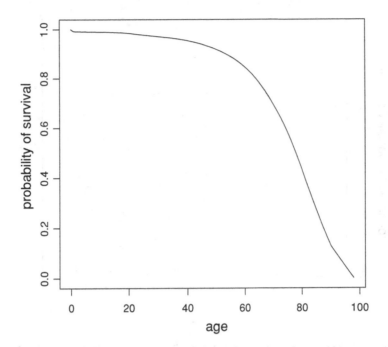

Figure 2.1. Life table survival curve (P_x-values)—U.S. white males. *Source:* National Center for Health Statistics, year 2000.

a slow decrease over the next 60 years; then, at about age 60, the survival probabilities begin to decline rapidly.

The conditional probabilities of death (the age-specific q_x-values) reflect the underlying life table hazard rate. For a life table one-year interval ($\delta = 1$), the approximate average mortality rate at age x is

$$\text{average mortality rate} = \frac{P_x - P_{x+1}}{\frac{1}{2}[P_x + P_{x+1}]} \qquad \text{(Chapter 1)}$$

and the approximate life table hazard rate [denoted h(x)] becomes

$$h(x) \approx \frac{P_x - P_{x+1}}{\frac{1}{2}[P_x + P_{x+1}]} = \frac{l_x/l_0 - l_{x+1}/l_0}{\frac{1}{2}[l_x/l_0 + l_{x+1}/l_0]} = \frac{d_x}{l_x - \frac{1}{2}d_x} \approx \frac{d_x}{l_x} = q_x$$

because for most ages, d_x is much less than l_x ($d_x \ll l_x$ because $l_x - (\frac{1}{2})d_x \approx l_x$). For life table mortality (Table 2.1) at age 60, $l_{60} = 84{,}539$ and $d_{60} = 1{,}070$, making $h(60) \approx q_{60} = 0.0127$. The approximate hazard function (a series of 92 connected q_x-probabilities) from the 2000 U.S. life table (Table 2.1) is displayed in Figure 2.2. A slight improvement over

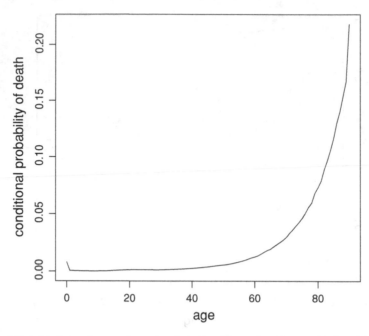

Figure 2.2.   Life table hazard curve (q_x-values)–U.S. white males. *Source:* National Center for Health Statistics, year 2000.

estimating the hazard rate by q_x is achieved by averaging two consecutive q_x-probabilities. That is, the life table hazard function is estimated by

$$h(x) \approx \tfrac{1}{2}(q_{x-1} + q_x).$$

As noted, the area under the survival curve equals the mean survival time. For a complete life table where each of the k intervals has length 1.0 $(t_x - t_{x-1} = 1)$ and $P_0 = 1, P_1, P_2, \ldots, P_{k-1}, P_k = 0$,

$$\text{mean lifetime} = \text{area} = \sum \tfrac{1}{2}(P_{x-1} + P_x) = \sum P_x - \tfrac{1}{2} = \sum \frac{l_x}{l_0} - \tfrac{1}{2}$$

$$= \frac{1}{l_0}\left[\sum\left(L_x + \tfrac{1}{2}d_x\right)\right] - \tfrac{1}{2}$$

$$= \frac{1}{l_0}\left[\sum L_x + \tfrac{1}{2}\sum d_x\right] - \tfrac{1}{2} = \frac{T_0 + \tfrac{1}{2}l_0}{l_0} - \tfrac{1}{2}$$

$$= \frac{T_0}{l_0} = e_0.$$

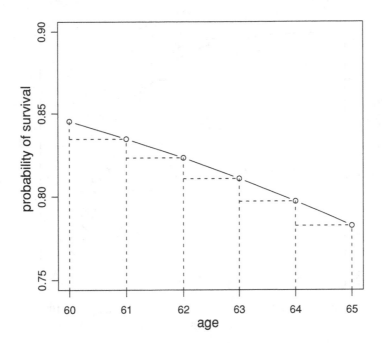

Figure 2.3.   Detail of the survival curve—U.S. white males.

Figure 2.3 displays a section of the life table survival curve in detail between ages 60 and 65, illustrating the geometric pattern (triangle + rectangle) associated with each one-year age interval. The mean lifetime calculated from the area under the life table survival curve is not exactly equal to the value e_0 from a current/complete life table (for example, Table 2.1) because the first ($x = 0$) and last age ($x = 90^+$) intervals have a different geometry.

Figure 2.4 displays the life table survival curves for both males and females based on the 2000 U.S. mortality rates. The areas under both survival curves reflect the mean lifetimes for each sex (area $\approx e_0$). For U.S. white males, the mean lifetime is $e_0 = 74.2$ years, and for U.S. females, the mean lifetime is $e_0 = 79.3$ years (U.S. mortality, 2000). The difference, $79.3 - 74.2 = 5.1$ years, is geometrically represented by the area between the survival curves (marked on the plot).

The 17th-century data collected by John Graunt to construct his original life table are still available and can be used to construct a modern life table. Of course, in Graunt's day there were no central statistical agencies and

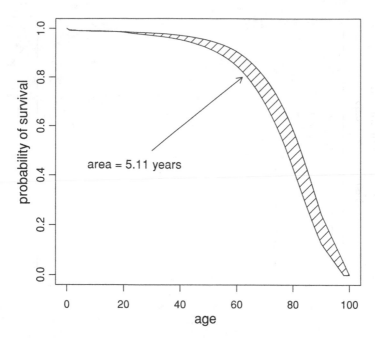

area = 5.11 years

Figure 2.4.   Difference between male and female 2000 U.S. life table survival curves. *Source:* National Center for Health Statistics, year 2000.

mortality data were not collected routinely. In fact, collecting mortality data was a new and innovative idea. Using primarily church records in the London area, he collected data consisting of dates of birth and death. These data are subject to a number of biases and do not represent anything near 100% of any population. Taken at face value, the original age-specific mortality rates produce the life table estimates shown in Table 2.3.

The survival curve (broken line) estimated from these data and the survival curve estimated from the 2000 U.S. population (solid line) are displayed in Figure 2.5. The considerable difference is clear. The extreme influence of 17th-century infant and childhood mortality is summarized by comparing the mean survival time, 18.9 years, to the modern U.S. value of 76.9 years (based on the 2000 U.S. life table). The difference of 58 years is again geometrically represented by the area between the two survival curves.

Graunt's life table data (Table 2.3) illustrate once again that the estimated mean survival time (e_0) is the area enclosed by the estimated survival curve. For each age interval x, the width is $t_x - t_{x-1} = \delta_x = 10$ years

Table 2.3. A modern life table constructed from John Graunt's 17th-century mortality data.

Age interval	δ_x	l_x	P_x	L_x	T_x	e_x
0–10	10	100	1.00	770	1890	18.9
10–20	10	54	0.54	440	1120	20.7
20–30	10	34	0.34	275	680	20.0
30–40	10	21	0.21	175	405	19.3
40–50	10	14	0.14	110	230	16.4
50–60	10	8	0.08	65	120	15.0
60–70	10	5	0.05	35	55	11.0
70–80	10	2	0.02	15	20	10.0
80–90	10	1	0.01	5	5	5.0

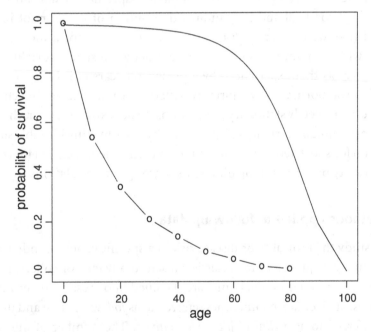

Figure 2.5.   Survival curve from John Graunt's 17th-century data (broken line) and U.S. 2000 survival curve (solid line).

and the area under the survival curve for each 10-year age interval is $\text{area}_x = \frac{1}{2}\delta_x(P_{x-1} + P_x)$, making the total area, as before,

$$e_0 = \text{area} = \sum \text{area}_x = \sum \frac{1}{2}\delta_x(P_{x-1} + P_x)$$
$$= \frac{1}{2}10(1.54 + 0.88 + 0.55 + \cdots + 0.03) = 18.9 \, \text{years}.$$

It is notable that the mean additional years of life at age 80 ($e_{80} = 10$ years) calculated from Graunt's 17th-century data is not very different from the mean found in the modern U.S. population ($e_{80} = 7.5$ years). The mean years of remaining life for extremely old individuals are frequently similar among diverse populations. A few e_{85}-values from selected male populations (year = 2000) are as follows: China = 3.9 years, Congo = 4.5 years, India = 4.1 years, Japan = 5.7 years, Mexico = 4.9 years, Norway = 5.7 years, and United States = 5.9 years.

A life table translates cross-sectional age-specific current mortality rates into a theoretical and longitudinal description of a cohort of individuals as if they were observed for more than 100 years. For a life table cohort, a newborn infant is assumed to have exactly the same mortality risk after 60 years as that experienced by a person 60 years old during the year the infant was born. Clearly, mortality patterns are not constant but change over time. Completely stationary populations do not exist. However, changes over time in human mortality risk are generally slow, making life table summaries useful for short-term predictions and, as mentioned, excellent for comparing mortality patterns among different subgroups or populations.

Life table methods applied to follow-up data

To study the mortality or disease risk in a specific group of individuals, frequently a sequence of intervals is chosen in advance of collecting the data and the observed survival times are classified into these predetermined categories. Such data are sometimes referred to as *follow-up data* and the analysis is referred to as a *clinical life table analysis*. The number of intervals chosen matters and typically influences the analytic results. Too few intervals provide insufficient detail to characterize the distribution of the data. Too many intervals reduce the number of observations in some or all of the intervals so that estimated values become unstable (large variances). Therefore,

Table 2.4. Follow-up data: tabulated hypothetical survival times ($n = 40$) displayed in Figure 2.6.

Intervals	d_i	w_i	$d_i + w_i$	l_i	l_i'	$\hat{q}_i$	$\hat{p}_i$	$\hat{P}_k$
0–1	2	9	11	40	35.5	0.056	0.944	0.944
1–2	2	6	8	29	26.0	0.077	0.923	0.871
2–3	4	1	5	21	20.5	0.195	0.805	0.701
3–4	3	3	6	16	14.5	0.207	0.793	0.556
4–5	2	1	3	10	9.5	0.211	0.789	0.439
5–6	2	1	3	7	6.5	0.308	0.692	0.304
6–7	0	0	0	4	4.0	0.000	1.000	0.304
7–8	1	3	4	4	2.5	0.400	0.600	0.182

classifying observations into a sequence of intervals (not just a follow-up table) involves a trade-off between detail and precision, requiring a bit of care to get the balance correct.

Consider $n = 40$ hypothetical follow-up times (in months):

1.2^+	5.0^+	0.3^+	3.0	1.3	0.9	7.2^+	2.3	3.4^+	2.7
2.8	1.6^+	1.1	1.1^+	0.7	3.9^+	1.7^+	7.3^+	4.5	7.5^+
1.2^+	0.9^+	0.6^+	0.2^+	2.1	2.1^+	5.0	4.0^+	0.8^+	5.0
0.5^+	1.8^+	3.6^+	0.1^+	7.9	4.2	0.1^+	3.4	0.4^+	$3.6.$

These data, for example, could be the survival times of 40 patients who received an experimental surgical treatment for a serious disease. The symbol "$+$" indicates an incomplete follow-up time. That is, when the study ended the treated person was still alive. When death occurred, it was at an unknown time after the study was concluded. These 40 survival times, classified into a sequence of eight one-month intervals, are presented in Table 2.4 and displayed in Figure 2.6.

Three kinds of individuals exist within each time interval, those who complete the interval alive (l_{i+1}), those who die during the interval (d_i), and those who did not complete the interval because of insufficient follow-up time (w_i), giving $l_{i+1} = l_i - d_i - w_i$. For example, consider the interval 3–4 months: of the 16 individuals who survived 3 months (began the interval $i = 4$), 10 individuals survived the entire interval (beyond 4 months), three died, and three were incomplete (denoted $l_3 = 16$, $d_3 = 3$, and $w_3 = 3$) making

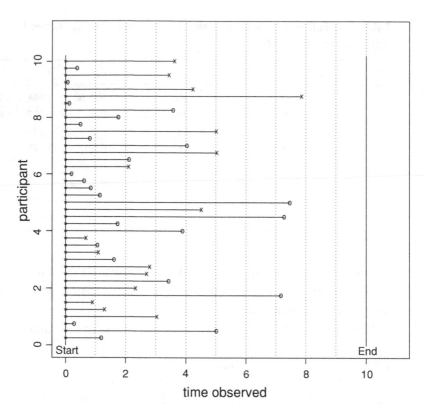

Figure 2.6.   Follow-up data: survival times of 40 hypothetical individuals from Table 2.4 (x = death and o = incomplete).

$l_4 = l_3 - d_3 - w_3 = 16 - 3 - 3 = 10$. Specifically, for the $l_3 = 16$ individuals who began interval 3 months, the survival times are

> died in the interval (3): 3.0, 3.4, and 3.6,
> incomplete in the interval (3): 3.4^+, 3.9^+, and 3.6^+,
> survived the entire interval (10): 5.0^+, 7.2^+, 7.3^+, 4.5, 7.5^+, 5.0, 4.0^+, 5.0, 7.9, and 4.2.

Individuals who complete an interval are observed for a time equal to the length of the interval. Individuals with incomplete survival times (time of death unknown) are observed for only part of the interval. It is reasonable to attribute a survival time to each these individuals equal to one-half the interval length. Consequently, they contribute to the total survival time (on average) half the survival time of the individuals who survive the entire interval. Based on this assumption, the number of persons alive at the beginning of the interval (l_i) is adjusted and called the *effective number of persons-at-risk*. The effective number of persons-at-risk (denoted l_i') is

effective number of persons-at-risk $= l_i' = l_i - \frac{1}{2}w_i$

for the ith interval, where l_i represents the number of individuals who began the interval and w_i represents the number of individuals who did not complete the interval. Using the effective number of persons-at-risk, the estimated conditional probability of death becomes $\hat{q}_i = d_i / l_i'$ (Table 2.4), compensating for the unobserved deaths. As previously, the estimated unconditional survival probability for the kth interval is

$$\hat{P}_k = \prod \hat{p}_i,$$

where $\hat{p}_i = 1 - \hat{q}_i$ and $i = 1, 2, \ldots, k$.

Another view of the effective number of individuals at risk shows explicitly the underlying assumption that all individuals in the same interval have the same probability of death. The observed number of deaths d_i is too small, because it does not include the unobserved deaths among the incomplete observations. The number of these individuals is effectively $0.5w_i$, because observing w_i individuals for one-half the interval is equivalent to observing $0.5w_i$ individuals for the whole interval. An estimate of the number of unobserved deaths among the incomplete observations (died after the ith interval ended and, therefore, not observed) becomes $0.5w_i\hat{q}_i$ for each interval. That is, an estimated $0.5w_i\hat{q}_i$ "missing deaths" for the ith interval would have been observed if complete survival times were known for all l_i individuals.

Increasing the observed number of deaths by the estimated amount $0.5w_i\hat{q}_i$ produces the expression for the probability of death for the ith interval,

$$\hat{q}_i = \frac{d_i + 0.5w_i\hat{q}_i}{l_i}.$$

Notice that the estimate of the probability $\hat{q}_i$ is the same for both incomplete and complete observations. Solving this expression for the conditional probability of death $\hat{q}_i$ gives the same expression as before,

$$\hat{q}_i = \frac{d_i}{l_i - \frac{1}{2}w_i} = \frac{d_i}{l_i'},$$

as long as the estimated value $\hat{q}_i$ applies equally to all sampled individuals.

Perhaps the simplest view of the number of effective persons-at-risk is the following: If all incomplete observations occurred immediately at the start of the interval, then the number of individuals at risk would be $l_i - w_i$, and if all incomplete observations occurred just before the end of the interval, then the number of individuals at risk would be l_i. The effective number of persons-at-risk is the mean of these two extreme conditions ($l_i' = l_i - 0.5w_i$). The use of the effective number of persons-at-risk to estimate q_i is sometimes called the *actuarial estimate*.

A common summary value calculated from follow-up data is called the "five-year survival rate." This "rate" is actually a five-year survival probability. For the hypothetical data, the estimated probability of surviving beyond five years is $\hat{P}_5 = 0.439$ (Table 2.4). An expression for the estimated variance, frequently called *Greenwood's variance*, produces an estimated variance of the distribution of the estimate $\hat{P}_i$ (details in Chapter 4). This estimated variance of the estimated survival probability $\hat{P}_5$ is variance ($\hat{P}_5$) = 0.011. The associated 95% confidence interval becomes approximately $0.439 \pm 1.960\sqrt{0.011}$ or (0.223, 0.655) based on the normal distribution and the estimated survival probability $\hat{P}_5 = 0.439$.

As with an incomplete observation, the complete survival time is not observed when an individual is lost from the study cohort. Frequently it is known in which interval an individual is lost. The effective "missing number of deaths" among the lost individuals is then estimated as $0.5u_i\hat{q}_i$, where u_i represents the number of lost individuals in the ith interval. For example, if six individuals are lost ($u_i = 6$), the effective "missing" number of deaths is estimated by $3\hat{q}_i$ for the entire interval. Including an estimate of the number of deaths among the lost individuals improves the estimate of the total number of deaths. In symbols, the estimated total number of deaths in the ith interval becomes $d_i + 0.5w_i\hat{q}_i + 0.5u_i\hat{q}_i$. The effective number of individuals at risk is then $l_i' = l_i - 0.5(w_i + u_i)$. Again, the effective number at risk is calculated as if the lost individuals had the same interval-specific probabilities of death as the observed individuals, namely the same probability q_i. Under these conditions, lost individuals are treated as incomplete observations.

Occasionally, individuals are lost from a study for reasons related to the outcome. For example, extremely sick individuals may no longer be able to continue to participate in the study, or relatively well individuals may be able to move away from the study area. Assuming that the probability of death for these kinds of lost individuals is the same as for those who remain in the study potentially biases the estimation of the survival probabilities $\hat{P}_k$. To get an idea of the extent of this bias, the estimates $\hat{P}_k$ can be calculated as if all lost individuals died. The estimated conditional probability of death becomes

$$\hat{q}_i' = \frac{d_i + 0.5u_i}{l_i - 0.5w_i}.$$

These theoretical probabilities ($\hat{q}_i$) yield the minimum possible survival probabilities ($\hat{P}_k'$) or the maximum bias that could occur from individuals lost to follow-up. The differences between the estimated survival probabilities $\hat{P}_k$ and $\hat{P}_k'$ indicate the largest possible bias that would occur by erroneously assuming that observed and lost individuals have the same probability of death. When all lost individuals are assumed to survive, the individuals are effectively not "lost" and the estimation of q_i does not need to be modified.

When sufficient numbers of survival times are sampled, follow-up data can be used to estimate a hazard rate (analogously to the previous life table estimate). A hazard rate is an instantaneous quantity but is pragmatically approximated by an estimated average rate over a short interval (Chapter 1). In symbols, for follow-up data,

$$\text{estimated hazard rate} = \hat{h}(t_i) = \frac{\text{number of deaths}}{\text{effective person-time at risk}} = \frac{d_i}{l_i'(t_i - t_{i-1})}$$

for the ith interval in a table such as Table 2.4. These estimates are the interval specific probabilities of death divided by the lengths of the associated time intervals. In symbols, for the ith interval, again, $\hat{h}(t_i) = \hat{q}_i/(t_i - t_{i-1})$. For the example (Table 2.5), the interval from 4 to 5 months ($i = 5$) yields the estimated hazard rate $\hat{h}(t_5) = 2/9.5 = 0.211$. Like a current life table, for the special case $t_i - t_{i-1} = 1$, the estimated hazard rate $\hat{h}(t_i)$ is again approximated by $\hat{q}_i$. Combining the hazard rates estimated for each time interval produces an estimate of the hazard function over the entire range of the survival times. The estimated hazard function for the hypothetical $n = 40$ survival times is displayed in Figure 2.7 based on the estimated values in Table 2.5, along with a smoothed estimate (dashed line).

Table 2.5. Follow-up data: estimated hazard function from the hypothetical 40 survival times (Table 2.4).

Intervals	d_i	l'_i	$\hat{h}_i(t)$	Std. error	Smoothed*
0–1	2	35.5	0.056	0.040	0.040
1–2	2	26.0	0.077	0.054	0.010
2–3	4	20.5	0.195	0.098	0.163
3–4	3	14.5	0.207	0.119	0.192
4–5	2	9.5	0.211	0.149	0.204
5–6	2	6.5	0.308	0.218	0.220
6–7	0	4.0	0.000	0.000	0.260
7–8	1	2.5	0.400	0.400	0.300

* See the appendix at the end of the chapter for a description of the smoothing process.

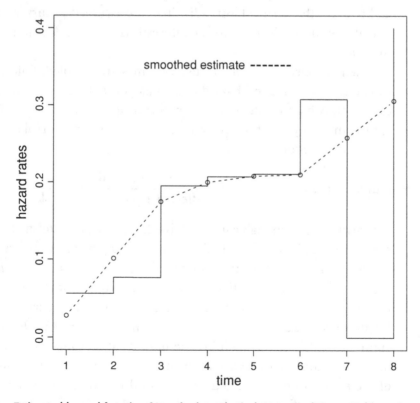

Figure 2.7. Estimated hazard function from the hypothetical 40 survival times (Table 2.5).

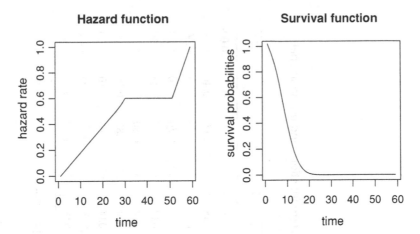

Figure 2.8. Comparison of a hazard function and a survival function applied to the same set of survival "data" (Table 2.6).

A slightly biased estimate of variance of the estimated hazard rate $\hat{h}(t_i)$ is given by the expression

$$\text{variance}[\hat{h}(t_i)] = \frac{[\hat{h}(t_i)]^2}{d_i}.$$

For most survival data, the interval-specific estimates of the hazard function are based on small numbers of observations (large standard errors—Table 2.5) and are frequently only rough approximations of the underlying hazard function (much more efficient estimates of a hazard function will be discussed, starting with Chapter 5).

One last point: Both the survival and hazard functions have important roles in the analysis of survival data. However, the hazard function is typically a more sensitive reflection of risk. From a descriptive point of view, the hazard function displays the risk of death or disease in a more intuitive fashion. To illustrate, the artificial data in Table 2.6 were created to emphasize the difference in descriptive properties of two summaries.

Figure 2.8 shows two descriptions of the same survival experience. Three different stages of risk are clearly identified by the hazard function and not by the survival function (a sharp increase in risk until about time $t = 30$, followed by a constant period until about time $= 50$, where increasing risk again occurs but at a high-rate than for the earlier pattern). The inability

Table 2.6. Data illustrating differences between conditional survival probabilities (hazard function) and unconditional survival probabilities of death (survival function) as a description of risk.

t_i	q_i	p_i	P_k	t_i	q_i	p_i	P_k	t_i	q_i	p_i	P_k	t_i	q_i	p_i	P_k
1	0.00	1.00	1.00	11	0.20	0.80	0.31	21	0.40	0.60	0.00	50	0.60	0.40	0.00
2	0.02	0.98	0.98	12	0.22	0.78	0.24	22	0.42	0.58	0.00	51	0.60	0.40	0.00
3	0.04	0.96	0.94	13	0.24	0.76	0.18	23	0.44	0.56	0.00	52	0.65	0.35	0.00
4	0.06	0.94	0.88	14	0.26	0.74	0.13	24	0.46	0.54	0.00	53	0.70	0.30	0.00
5	0.08	0.92	0.81	15	0.28	0.72	0.10	25	0.48	0.52	0.00	54	0.75	0.25	0.00
6	0.10	0.90	0.73	16	0.30	0.70	0.07	26	0.50	0.50	0.00	55	0.80	0.20	0.00
7	0.12	0.88	0.64	17	0.32	0.68	0.05	30	0.60	0.40	0.00	56	0.85	0.15	0.00
8	0.14	0.86	0.55	18	0.34	0.66	0.03	35	0.60	0.40	0.00	57	0.90	0.10	0.00
9	0.16	0.84	0.47	19	0.36	0.64	0.02	40	0.60	0.40	0.00	58	0.95	0.05	0.00
10	0.18	0.82	0.38	20	0.38	0.62	0.01	45	0.60	0.40	0.00	59	1.00	0.00	0.00

Note: Purely fictional data designed to dramatically illustrate the differences in the descriptive properties.

to clearly identify patterns of risk with a survival function becomes an issue because major changes in survival frequently take place after the survival probabilities have become small, making these changes difficult to detect. In addition, the period of lowest survival (highest risk) is frequently the most informative part of a survival function. In the example, by time $t = 20$ all remaining survival probabilities are less than 0.01 and these changes in risk are not visible on the survival probability scale.

APPENDIX

Statistical tools: a smoothing technique

Most smoothing techniques are variations on a simple principle. A sequence of n values $\{x_1, x_2, \ldots, x_n\}$ is made smooth by increasing the similarity of neighboring values. There are many ways to make neighboring values similar and many ways of defining neighborhood. The following describes one smoothing technique among a large variety of choices that is an easily applied and usually an effective approach.

The first step (called the *median moving average*) consists of choosing a value and the next smallest and next largest neighbors, and then replacing the chosen value with the median of these three values. In symbols,

$$\text{step 1: } x_i' = \text{median} (x_{i-1}, x_i, x_{i+1}) \qquad \text{for } i = 2, 3, \ldots, n - 1.$$

This median moving average removes extreme values, replacing them with locally similar values. However, the process tends to produce sequences of identical values (level spots on a plot) and has no influence on a sequence of three strictly increasing values.

A second step consists of a *mean moving average*. For this step, the moving average is a weighted sum of three consecutive values. Each already smoothed value (x_i') is replaced by a moving average of three consecutive values $(x_{i-1}', x_i', x_{i+1}')$. In symbols,

$$\text{step 2: } x_i^* = 0.25x_{i-1}^* + 0.50x_i' + 0.25x_{i+1}' \qquad \text{for } i = 2, 3, \ldots, n - 1.$$

The moving average removes "flat" sequences and smooths sequences of increasing values. The first and last values are not affected by the two kinds of moving averages. Using the twice-smoothed sequence to predict these two values yields

$$x_1^* = 2x_2^* - x_3^* \quad \text{(first value)} \quad \text{and} \quad x_n^* = 2x_{n-1}^* - x_{n-2}^* \quad \text{(last value)}.$$

The following table illustrates this three-step smoothing technique applied to the eight estimated hazard rates in Table 2.5 and displayed in Figure 2.7:

i	1	2	3	4	5	6	7	8
$x_i = \hat{h}(t_i)$	0.06	0.08	0.20	0.21	0.21	0.31	0.00	0.40
step 1 (x'_i)	0.06	0.08	0.20	0.21	0.21	0.21	0.21	0.40
step 2 (x_i^*)	0.06	0.10	0.16	0.19	0.20	0.22	0.26	0.40
smoothed	0.04	0.10	0.16	0.19	0.20	0.22	0.26	0.30

Two especially useful estimation tools

Estimates of parameters based on statistical models and their evaluation are major components of statistical methods. The following outlines two techniques that are key to statistical estimation in general, namely maximum likelihood estimation and the derivation of the statistical properties of analytic functions. These somewhat theoretical topics are not critical to understanding the application of survival analysis methods, but provide valuable insight into the origins of parameter estimates and the variances of their distributions.

Maximum likelihood estimation

Maximum likelihood estimation is used in the vast majority of statistical analyses to determine values for the parameters of models describing the relationships within sampled data. The complexity of this technique lies in the technical application and not its underlying principle. Maximum likelihood estimation is conceptually simple. A small example introduces the fundamental considerations at the heart of the maximum likelihood estimation process.

Suppose that a thumbtack tossed in the air has an unknown probability of landing with the point up (denoted p). Furthermore, three tacks are tossed and one lands point up and the other two land point down. The probability of this result is $3p(1 - p)^2$. When two values are proposed as an estimate of p, it is not hard to decide the most likely to have produced the observed result (one up and two down) and therefore is the better estimate of the unknown probability p. For the example, the likelihood that one up-tack occurs out of three tossed when $p = 0.2$ is $3(0.2)(0.8)^2 = 0.384$ and the probability of the same outcome when $p = 0.8$ is $3(0.8)(0.2)^2 = 0.096$. The "maximum likelihood" question becomes: Which of the two postulated probabilities (0.2 or 0.8) is the better estimate of the unknown underlying value p? Although no unequivocal answer exists, selecting the probability $p = 0.2$ is simply more sensible. The observed data "better" support this answer because $p = 0.2$ makes the observed results four times more likely than $p = 0.8$. Of the two choices, selecting $p = 0.2$ is more consistent with the observed data.

Maximum likelihood estimation is an extension of the same logic. The data are considered as fixed and all values of the parameter are considered (not just two values). The parameter that makes the observed data the most likely (maximizes the likelihood of their occurrence) is chosen as the "best" estimate. It is again the value most consistent with the observed data. For the thumbtack example, this value is 0.333. No other choice of p makes the observed data (one up and two down) more likely. For all other possible values of the parameter p, the probability $3p(1 - p)^2$ is less than $3(0.333)(0.667)^2 = 0.444$. For the tack data (one-up and two-down), the maximum likelihood estimate $\hat{p} = 0.333$ is not the correct value but is the most sensible choice in light of the observed data.

Expanding this example continues to illustrate the logic of the maximum likelihood estimation process. Say $n = 50$ tacks are tossed in the air and $x = 15$ land point up. That is, the data are

up, up, down, up, down, down, ..., down, and up.

The probability that this event occurred is

$$L = p \times p \times (1 - p) \times p \times (1 - p) \times (1 - p) \times \cdots \times (1 - p) \times p$$

or more succinctly

$$L = p^{15}(1 - p)^{35}.$$

The expression L is called the *likelihood function*. As with the previous examples, the value of the parameter p is unknown. The maximum likelihood question becomes, Out of all possible values for p, which value makes the observed result ($x = 15$ up-tacks) most likely to have occurred? The answer is found by calculating the likelihood function L for all possible values of p and identifying the largest value. Because sums are easier to describe conceptually and deal with mathematically, instead of the likelihood L (a product), the logarithm of L (a sum) is used [denoted $\log(L)$]. For the thumbtack example,

Table 3.1. Selected values of the parameter p and the corresponding log-likelihood values for $x = 15$ up-tacks among $n = 50$ tosses.

p	0.05	0.10	0.15	0.20	0.25	0.30	0.35	0.40	0.45	0.50	0.55	0.60
$\log(L)$	−46.7	−38.2	−34.1	−32.0	−30.9	−30.5	−30.8	−31.6	−32.9	−34.7	−36.9	−39.7

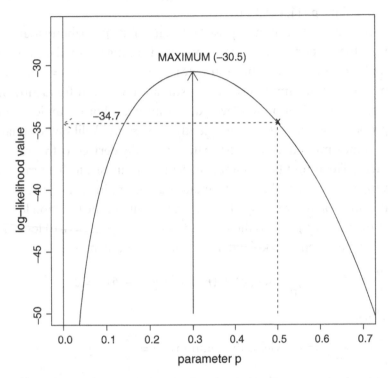

Figure 3.1. The log-likelihood function for the thumbtack data ($n = 50$ and $x = 15$).

the likelihood value L is the product $L = p^{15}(1 - p)^{35}$ and the log-likelihood value is the sum $\log(L) = 15\log(p) + 35\log(1 - p)$. The value that maximizes the log-likelihood function also maximizes the likelihood function. For the thumbtack data, 12 selected values of p produce the log-likelihood values in Table 3.1. In fact, the possible values of p range continuously from 0 to 1. Figure 3.1 displays the log-likelihood values $\log(L)$ for a relevant range of p (0.04 to 0.7).

The log-likelihood function increases until $p = 0.3$ and then decreases. The value 0.3 is the value of p that maximizes the log-likelihood function

$\log(L)$ and, therefore, maximizes the likelihood function L. It is denoted $\hat{p}$ and called *the maximum likelihood estimate of the parameter* p. To repeat, no other value is more consistent with the observed data. The occurrence of 15 up-tacks (35 down-tacks) is most likely when p is 0.3, making $\hat{p} = 0.3$ the maximum likelihood estimate. Technically, the log-likelihood value $\log(L_{p=03}) = -30.5$ is greater than $\log(L)$ for all other possible values of the probability p (Figure 3.1).

A natural and commonly used estimate of the probability that a tack lands up is the proportion of up-tacks observed among the total number tossed or, for the example, the natural estimate is $\hat{p} = x/n = 15/50 = 0.3$. An amazing property of maximum likelihood estimation is that it frequently provides a rigorous justification for "everyday" estimates. For example, mean values, proportions, and rates are frequently maximum likelihood estimates.

A maximum likelihood estimate is typically derived with a calculus argument. The thumbtack example continues to illustrate. The maximum of a single-valued log-likelihood function is that point where the derivative of the function is zero. In symbols, the maximum of the function $\log(L)$ occurs at the value of p that is the solution to the equation $(d/dp)\log(L) = 0$. For example, when x tacks land up out of n tosses, then

$$\frac{d}{dp}\log(L) = \frac{d}{dp}[x\log(p) + (n-x)\log(1-p)] = 0.$$

Thus,

$$\frac{x}{\hat{p}} - \frac{n-x}{1-\hat{p}} = 0 \quad \text{yields the solution} \quad \hat{p} = \frac{x}{n},$$

where $\log(L) = \log[p^x(1-p)^{n-x}] = x\log(p) + (n-x)\log(1-p)$ is the log-likelihood function for all possible parameter values p $(0 \le p \le 1)$. Again, the estimated value $\hat{p} = x/n$ maximizes the likelihood function and is also the natural estimate of the parameter p (proportion of up-tacks).

In addition, the variance of a maximum likelihood estimate can be estimated from the log-likelihood function. For the example, the variance of the distribution of the estimate $\hat{p}$ is estimated by $\hat{p}(1-\hat{p})/n$. In general, maximum likelihood estimates are found with a computer program, so the details of the numerical estimation process and the derivation of the variance expressions are rarely issues in analyzing data and are left to more theoretical presentations.

When more than one parameter is estimated, the notation and computation become more elaborate, but the maximum likelihood principle remains the same. Regardless of the complexity of the likelihood function, the estimates are the set of values that are most likely to have produced the observed data. Suppose that l parameters are to be estimated; then the maximum likelihood estimates are the l parameter values that make the likelihood function the largest possible. In symbols, the l parameters denoted $\theta_1, \theta_2, \theta_3, \ldots, \theta_l$ have maximum likelihood estimates $\hat{\theta}_1, \hat{\theta}_2, \hat{\theta}_3, \ldots, \hat{\theta}_l$ when the likelihood value L evaluated at these values is larger than the likelihood values calculated from all other possible parameter values $\theta_1, \theta_2, \theta_3, \ldots, \theta_l$ or $L(\hat{\theta}_1, \hat{\theta}_2, \hat{\theta}_3, \ldots, \hat{\theta}_l) > L(\theta_1, \theta_2, \ldots, \theta_l)$. The computer techniques applied to find this set of estimates are sophisticated and complex but the interpretation of the estimated values remains simple. They are the unique set of estimates that are most consistent with the observed data. In other words, among all possible sets of the parameters, it is that set that makes the occurrence of the observed data most likely.

For most statistical techniques, the parameter values are thought of as fixed and the data are subject to sampling variation. Maximum likelihood estimation reverses the situation. The observed data are considered fixed and the parameters are varied over all possible values to determine the specific value or values that maximize the likelihood function.

It is frequently difficult to construct the likelihood function L for a statistical model. Furthermore, the computational process necessary to find specific estimates and their variances is extremely tedious for more than two parameters. Consequently, a computer algorithm is almost always used to create the likelihood function and to estimate statistical model parameters.

Four key properties of maximum likelihood estimates

1. Maximum likelihood estimates based on large numbers of observations have approximate normal distributions. Often, as few as 10 or 20 observations are sufficient to produce estimates with approximately normal distributions. Therefore, the assessment of maximum likelihood estimates in terms of confidence intervals and statistical tests follows typical patterns. For example, when $\hat{\theta}$ represents a maximum likelihood parameter estimate, an approximate 95% confidence interval is $\hat{\theta} \pm 1.960 \sqrt{\text{variance}(\hat{\theta})}$

based on the approximate normal distribution. In addition, the test statistic

$$X^2 = z^2 = \frac{(\hat{\theta} - \theta_0)^2}{variance(\hat{\theta})}$$

has an approximate chi-square distribution with one degree of freedom when θ_0 is the "true" underlying parameter value estimated by $\hat{\theta}$. The word "true" in this statistical context means that the difference between the estimate $\hat{\theta}$ and the parameter θ_0 is due entirely to random variation. The maximum likelihood estimated variance, denoted variance($\hat{\theta}$), serves as an estimate of the variance of the approximate normal distribution that describes the variation of the estimated value, $\hat{\theta}$. This chi-square assessment of a maximum likelihood estimated parameter is sometimes called *Wald's test*.

2. A maximum likelihood estimate is optimal in the sense that it usually has a smaller variance than competing estimates. When the sample size is large, the maximum likelihood estimate is always the most precise estimate available (smallest variance). Thus, for a wide variety of analyses, the estimates that most efficiently utilize the sampled data are found by maximizing the likelihood function.

3. As noted, the estimated variance of a maximum likelihood estimate is necessarily calculated as part of the estimation process. The computer algorithm that produces the estimates produces estimates of their variances.

4. A function of a maximum likelihood estimate is itself a maximum likelihood estimate and has properties 1, 2, and 3. When the estimate $\hat{\theta}$ represents a maximum likelihood estimate, then, for example, $e^{\hat{\theta}}$ or $\sqrt{\hat{\theta}}$ or $n\hat{\theta}$ or $1/\hat{\theta}$ is also a maximum likelihood estimate. These estimates also have minimum variance and approximate normal distributions for large sample sizes. For the thumbtack tossing example, because $\hat{p} = 0.3$ is the maximum likelihood estimate of the probability that a tack lands up, then $n\hat{p} = 100(0.3) = 30$ is the maximum likelihood estimate of the number of up-tacks that would occur among $n = 100$ tosses. Furthermore, the probability $\hat{q} = 1 - \hat{p} = 1 - 0.3 = 0.7$ makes $\hat{q}$ the maximum likelihood estimate of the probability that a tack lands point down.

Likelihood statistics

Producing optimal parameter estimates and their estimated variances from sampled data is only one of the valuable properties of a likelihood function. The likelihood function or the logarithm of the likelihood function (not just the maximum) reflect the probability that the collected data occurred for a specific set of parameters. For example, if $p = 0.8$, then the likelihood of one-up and two-down tacks is $L = 3p(1 - p)^2 = 3(0.8)(0.2)^2 = 0.096$ ($\log(L) = -2.343$).

Table 3.1 shows that the log-likelihood value $\log(L)$ is -34.7 when $p = 0.5$ is postulated as the underlying probability for tossing $n = 50$ thumbtacks. This likelihood value is clearly not the maximum but, nevertheless, reflects the probability that fifteen up-tacks occur as if p were 0.5. The maximum value of the log-likelihood occurs at $\hat{p} = 0.3$ and is $\log(L) = -30.5$.

The two log-likelihood values -34.7 and -30.5 differ for one of two distinct reasons. Because the estimate $\hat{p}$ is subject to random variation, the two log-likelihood values possibly differ simply by chance alone when $p = 0.5$ is the underlying parameter. Alternatively, the value $p = 0.5$ may not be the underlying parameter, causing the two log-likelihood values to differ systematically. The question of why $\hat{p} = 0.3$ differs from $p = 0.5$ is addressed by the difference in log-likelihood values. The larger the difference, the smaller the probability that the two likelihood values differ by chance alone.

To help choose between these two alternatives (random versus systematic?) based on the observed difference between two log-likelihood values generated from two statistical models, a theorem from theoretical statistics is an important analytic tool. The theorem states that the difference between two log-likelihood values, multiplied by -2, has an approximate chi-square distribution when three conditions hold. The first condition is that the two models generating the log-likelihood values must be calculated from exactly the same data. Second, the compared models must be nested. Nested means that one model is a special case of the other (examples follow). Third, the two log-likelihood values must differ only because of random variation. When the first two conditions apply, a test statistic with a chi-square distribution (called the *likelihood ratio test statistic*) produces an assessment of the plausibility of the third condition in terms of a probability. The question becomes: Is the observed difference between log-likelihood values calculated from the same data an indication of systematic differences between two nested models or likely due only to the influence of random variation? To help answer this question, the comparison of two log-likelihood values and a chi-square distribution produces a significance probability (p-value). Specifically, the likelihood ratio test statistic is

$$X^2 = -2[\log(L_0) - \log(L_1)]$$

and has an approximate chi-square distribution with m degrees of freedom when $\log(L_0)$ and $\log(L_1)$ calculated from two nested models differ by chance alone. The degrees of freedom m is the difference between the number of parameters estimated to calculate each log-likelihood value.

The thumbtack tossing example conforms to all three requirements if the underlying underlying probability that a tack lands up is $p = 0.5$ (null hypothesis) and the estimated value $\hat{p} = 0.3$ differs strictly by chance alone. Then, the likelihood ratio test statistic (using the log-likelihood values from Table 3.1 and Figure 3.1)

$$X^2 = -2[\log(L_{p=0.5}) - \log(L_{p=0.3})] = -2[(-34.657) - (-30.543)] = 8.228$$

is a single observation from a chi-square distribution with one degree of freedom. For $\log(L_{p=0.3})$, one estimate is made (namely, $\hat{p} = 0.3$), and for $\log(L_{p=0.5})$, no estimate is made ($p = 0.5$ was selected), yielding one degree of freedom. The probability that a more extreme difference between the log-likelihood values occurs by chance alone is then p-value $= P(X^2 \geq 8.228 \mid p = 0.5) = 0.004$ from a chi-square distribution with one degree of freedom. Thus, the actual value of the parameter p is not likely in the neighborhood of 0.5 but is likely closer to 0.3. The conjecture that the selected value of $p = 0.5$ is the underlying parameter of the distribution that produced the thumbtack data and the estimate $\hat{p} = 0.3$ occurred by chance is not plausible.

In fact, the comparison of two likelihood values produces statistical tests similar to many familiar procedures. Such likelihood comparisons give results similar to chi-square tests, t-tests, and tests based on the normal distribution, such as approximate tests of a proportion. For example, when again $n = 50$

and $x = 15$, the classic test of the hypothesis of $p = p_0$ is based on the test statistic

$$z = \frac{\hat{p} - p_0}{\sqrt{\frac{\hat{p}(1 - \hat{p})}{n}}}$$

and $z^2 \approx X^2$ has an approximate chi-square distribution with one degree of freedom. Specifically, for the thumbtack example and $p_0 = 0.5$,

$$z = \frac{0.3 - 0.5}{\sqrt{\frac{0.3(1 - 0.3)}{50}}} = -3.086.$$

The test statistic is then $z^2 = (-3.086)^2 = 9.524$ and is approximately equal to the likelihood chi-square value of $X^2 = 8.228$. The difference between these two approaches diminishes as the sample size increases or estimates become closer to the hypothesis-generated test values. For example, when $p_0 = 0.4$, then $z^2 = (-1.543)^2 = 2.381$ and the likelihood ratio chi-square test statistic is $X^2 = 2.160$.

Calculating the difference between log-likelihood values is a fundamental statistical tool and applies to comparing analytic models with any number of parameters. In general, for a model containing k variables and l parameters, the log-likelihood value is represented by

$$\log(L_1) = \text{log-likelihood} = \log[L(\theta_1, \theta_2, \theta_3, \ldots, \theta_l \mid x_1, x_2, x_3, \ldots, x_k)].$$

Notice that the likelihood value remains conditional on the observed data (as if the data were fixed). A second log-likelihood value based on creating a nested model by removing m parameters (set equal to zero) is represented by

$$\log(L_0) = \text{log-likelihood}$$
$$= \log[L(\theta_1 = 0, \ldots, \theta_m = 0, \theta_{m+1}, \ldots, \theta_l \mid x_1, x_2, x_3, \ldots, x_k)].$$

Or, for a model based on the remaining $l - m$ parameters, the same likelihood value is represented as

$$\log(L_0) = \text{log-likelihood} = \log[L(\theta_{m+1}, \ldots, \theta_l \mid x_1, x_2, x_3, \ldots, x_k)].$$

As long as these two log-likelihood values are calculated from the same data, contrast nested models, and differ only because of random variation, the likelihood ratio test statistic $X^2 = -2[\log(L)_0 - \log(L_1)]$ has a chi-square distribution. The degrees of freedom are m, where m represents the number of parameters deleted from the more complex model (most parameters) to form the simpler and, as required, nested model.

The comparison of two log-likelihood values (almost always calculated with a computer program) reflects the difference in relative "fit" between two sets of conditions described in term of two nested models. The observed difference indicates the effectiveness of the simpler model (based on fewer parameters) in summarizing the observed data, compared to that of the more complex model. When a parameter value or a set of parameter values is eliminated from a model and the log-likelihood value remains essentially unaffected (only a slight decrease), the inference is made that the parameters eliminated are relatively unimportant and likely have only random influences. The log-likelihood difference is said to be consistent with random variation. Conversely, when a parameter value or a set of parameter values is eliminated from a model and the log-likelihood value decreases strikingly, the inference is made that the parameters eliminated are important and likely have systematic influences. The comparison of two log-likelihood values, therefore, produces a chi-square test that allows an evaluation of the difference between two models induced by eliminating selected model parameters (random or systematic?) in terms of a significance probability.

Example

The following illustrates a typical contrast of two nested models. Using the estimated parameters $\hat{b}_i$ to generate two log-likelihood values, the linear model $y = a + b_1 x_1 + b_2 x_2 + b_3 x_3$ is compared to the nested model $y = a + b_1 x_1$ to evaluate the role of the variables represented by x_2 and $x_3 (b_2 = b_3 = 0?)$. The likelihood ratio chi-squared test statistic $X^2 = -2[\log(L_0) - \log(L_1)] = 6.2$ (last column, Table 3.2) has an approximate chi-square distribution with two degrees of freedom $(m = 2)$, producing the p-value $P(X^2 \geq 6.2 \mid b_2 = b_3 = 0) = 0.045$, when x_2 and x_3 have only random influences on the variable y. The moderately small p-value gives some indication that parameter b_2 or b_3 or both are important components of the more complex model.

A log-likelihood value by itself is frequently not a useful assessment of the goodness-of-fit (a comparison between model-generated values and the

Table 3.2. The evaluation of the likelihood ratio statistic for two nested models (hypothetical).

Nested models	Constraint	d.f.	Log-likelihood values
$a + b_1x_1 + b_2x_2 + b_3x_3$	—	$n - 4$	$\log(L_1) = -145.3$
$a + b_1x_1$	$b_2 = b_3 = 0$	$n - 2$	$\log(L_0) = -148.4$
Difference	—	$m = 2$	$-2[\log(L_0) - \log(L_1)] = 6.2$

data). The absolute magnitude of a log-likelihood value is primarily determined by the sample size; the larger the sample size, the smaller the log-likelihood statistic. Because a difference between two likelihood values from nested models is not influenced by the sample size (same data for both calculations), it reflects the relative difference between the compared models. The important issue of the adequacy of the model to represent the relationships within the data is frequently addressed with other methods.

A sometimes handy rule of thumb states that when the likelihood ratio chi-square test statistic X^2 is less than m (the number of parameters eliminated), no evidence exists that these parameters play a systematic role in the model. The rule is simply an application of the fact that the mean value of a chi-square distribution is its degrees of freedom. The likelihood ratio test-statistic has a chi-square distribution with m degrees of freedom when the m eliminated parameters have only random influences. Therefore, an observed chi-square value less than its mean value m ($X^2 < m$) is not extreme, providing no evidence of a systematic difference between likelihood values. The smallest possible p-value is always greater than 0.3. Of course, exact probabilities exist in tables or are part of statistical computer programs.

The statistical properties of the function $f(x)$

Consider a variable denoted x with known or postulated properties but questions arise concerning a function of x, denoted $f(x)$. Two important statistical questions are: What is the mean and what is the variance of the distribution of $f(x)$? Or, in symbols,

mean of the distribution of $f(x) = ?$ and   variance$[f(x)] = ?$

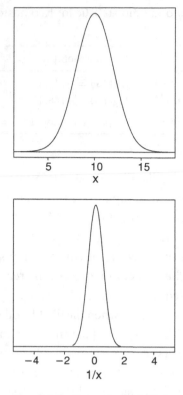

Figure 3.2.   The distributions of x and $f(x) = 1/x$.

Two rules allow the mean and the variance of the distribution of the variable $f(x)$ to be derived from the mean and variance of the distribution of the variable x. They are as follows:

Rule 1. The mean of the distribution of the variable $f(x)$ (denoted μ_f) is approximately the value of the function evaluated at the mean of the distribution of x (denoted μ). That is,

μ_f = mean of the distribution of $f(x) \approx f(\mu)$,

where μ_f represents the approximate mean value of the distribution of the $f(x)$-values and μ represents the mean value of the distribution of the x-values.

Rule 2. The variance of the distribution of the variable $f(x)$ (denoted variance$[f(x)]$) is approximately

$$\text{variance of } f(x) = \text{variance } [f(x)] \approx \left[\frac{d}{dx} f(\mu)\right]^2 \text{variance } (x).$$

The symbol $(d/dx) f(\mu)$ represents the derivative of the function $f(x)$ with respect to x, evaluated at the mean value μ.

Both rules 1 and 2 are an application of a basic mathematical tool called a *Taylor series expansion* [2], and the application to a statistical function is sometimes referred to as "the delta method."

For example, suppose the variable x has a symmetric distribution (Figure 3.2, top) with a mean value $= \mu = 10$ and a variance$(x) = 4$. The distribution of $f(x) = 1/x$ (Figure 3.2, bottom) then has an approximate mean value $\mu_f = 1/\mu = 1/10 = 0.1$ (rule 1) and an approximate variance$[f(x)] = (1/\mu^4)$variance$(x) = (1/10^4)4 = 0.0004$ (rule 2) because the derivative of $1/x$ is

$$\frac{d}{dx}f(x) = \frac{d}{dx}\left[\frac{1}{x}\right] = -\frac{1}{x^2} \quad \text{and} \quad \left[\frac{d}{dx}f(\mu)\right]^2 = \frac{1}{\mu^4} = \frac{1}{10^4}.$$

Application 1

A Poisson-distributed variable is sometimes transformed by taking the square root to produce a more normal-like distribution. For a Poisson distribution, the mean value (represented by $\mu = \lambda$) and the variance (represented by variance$(x) = \lambda$) are equal. The function $f(x) = \sqrt{x}$ produces a more symmetric and approximate normal distribution (if λ is not too small) with mean value $= \sqrt{\lambda}$ and variance $= 1/4$.

Applying rules 1 and 2, the approximate mean value of the normal-like distribution of $\sqrt{x}$ is

$$\mu_f = \text{mean of the distribution of } \sqrt{x} = \mu_{\sqrt{x}} \approx \sqrt{\mu} = \sqrt{\lambda} \text{ (rule 1)}.$$

The approximate variance is

$$\text{variance of } f(x) = \text{variance}(\sqrt{x}) = \frac{1}{4\lambda} \text{ variance}(x) = \frac{1}{4\lambda}\lambda = \frac{1}{4} \text{ (rule 2)}$$

because the derivative of $\sqrt{x}$ is

$$\frac{d}{dx}f(x) = \frac{d}{dx}\sqrt{x} = \frac{1}{2\sqrt{x}} \quad \text{and} \quad \left[\frac{d}{dx}f(\lambda)\right]^2 \approx \frac{1}{4\lambda}.$$

Application 2

Applying rules 1 and 2 to the logarithm of a variable x again yields an expression for the approximate mean value and variance of the distribution of the transformed variable $\log(x)$. Thus, when $f(x) = \log(x)$, applying rule 1 yields

$$\mu_f = \text{mean of the distribution of } f(x) = \mu_{\log(x)} \approx \log(\mu),$$

where μ represents the mean of the distribution of the variable x. The derivative of $\log(x)$ is

$$\frac{d}{dx} f(x) = \frac{d}{dx} \log(x) = \frac{1}{x}$$

and applying rule 2 yields

$$\text{variance of } f(x) = \text{variance}[f(x)] = \text{variance}[\log(x)] \approx \frac{1}{\mu^2} \text{variance}(x).$$

Corollary

$$\text{variance}(x) \approx \mu^2 \text{variance}[\log(x)].$$

Statistical summaries and estimates are frequently transformed to have more symmetric (normal-like) distributions using logarithms. Such transformations frequently improve the accuracy of statistical tests and confidence intervals. The mean value and variance of the distribution of the logarithm of a variable then become necessary parts of the statistical evaluation.

Application 3

The variance of the logarithm of a count is frequently estimated by the reciprocal of the count. That is, when a count (denoted m) is observed, the estimated variance of $\log(m)$ is $\text{variance}(\log[m]) = 1/m$. Such reciprocal values appear in the variances of the logarithm of odds ratios, the variance of the logarithm of odds (log-odds), and the variance of the logarithm of rate-ratios. The estimated variance of the function $f(m) = \log(m)$ results from applying rule 2.

When the count m has a binomial distribution,

$$\text{the mean} = \mu = np \quad \text{and} \quad \text{variance}(m) = np(1 - p).$$

When the count m has a Poisson distribution

$$\text{the mean} = \mu = \lambda = np \quad \text{and} \quad \text{variance}(m) = \lambda = np.$$

Therefore, when p is small $(1 - p \approx 1)$ or when m has a Poisson distribution,

$$\text{variance}(\log[m]) \approx \left[\frac{-1}{np}\right]^2 np = \frac{1}{np} \quad \text{(rule 2)}$$

because $\dfrac{d}{dm} f(m) = \dfrac{d}{dm} \log(m) = \dfrac{1}{m}$ and $\left[\dfrac{d}{dm} f(\mu)\right]^2 = \left[\dfrac{-1}{np}\right]^2$.

When m events occur among a series of n observations, the probability of the occurrence of m is estimated by $\hat{p} = m/n$. Therefore, a natural estimate of the variance of the logarithm of m is

$$S^2_{\log(m)} = \frac{1}{n\hat{p}} = \frac{1}{m}$$

because $n\hat{p} = m$.

Product-limit estimation

Summary values are at the heart of statistical analysis. Similarly, the estimation of survival probabilities and hazard rates is at the center of summarizing survival data. As might be imagined, a number of ways exist to estimate these two fundamental statistical summaries of survival experience.

To start, suppose a sample consists of n independent, unique, and complete survival times. Unique means that all sampled survival times are different. Complete means that all survival times end in an observed outcome such as death. For example, consider the unique and complete survival times of 10 ($n = 10$) extremely ill AIDS patients:

survival times (in days):   2, 72, 51, 60, 33, 27, 14, 24, 4, and 21.

A simple and direct estimate of a survival probability begins with calculating the conditional probability that an individual dies within a specific time interval. Among several ways to create these intervals, one is based on the time of death. Each interval is defined so that it contains only one death. In this case, the interval limits are constructed from survival times (denoted t_i). Unlike constructing a life table, the interval lengths vary and are determined by observed values. For the example AIDS data, a sequence of 10 such intervals (denoted t_{i-1} to t_i) is given in Table 4.1 (second column). For example, the sixth interval t_5 to t_6 is 24 days to 27 days ($i = 6$) and, like all 10 intervals, contains one death.

For these complete data, the estimated conditional probability of dying within a specific interval (ith interval) is

$$\hat{q}_i = P(\text{death before time } t_i \mid \text{alive at } t_{i-1}) = \frac{\text{number of deaths}}{\text{number at risk}}$$

$$= \frac{1}{n - i + 1}$$

Table 4.1. Interval specific conditional survival probabilities of death for the AIDS data ($n = 10$ unique and complete survival times).

i	Intervals $t_{i-1} - t_i$	Deaths d_i	At-risk $n - i + 1$	Probabilities $\hat{q}_i$	Probabilities $\hat{p}_i$
1	0–2	1	10	1/10	$9/10 = 0.900$
2	2–4	1	9	1/9	$8/9 = 0.889$
3	4–14	1	8	1/8	$7/8 = 0.875$
4	14–21	1	7	1/7	$6/7 = 0.857$
5	21–24	1	6	1/6	$5/6 = 0.833$
6	24–27	1	5	1/5	$4/5 = 0.800$
7	27–33	1	4	1/4	$3/4 = 0.750$
8	33–51	1	3	1/3	$2/3 = 0.667$
9	51–60	1	2	1/2	$1/2 = 0.500$
10	60–72	1	1	1/1	$0/1 = 0.000$

and the estimated conditional probability of surviving the entire interval (surviving from t_{i-1} to t_i) is

$$\hat{p}_i = 1 - \hat{q}_i = P(\text{alive after time } t_i \mid \text{alive at } t_{i-1}) = \frac{n - i}{n - i + 1}.$$

The denominator $n_i = n - i + 1$ is the number of individuals at risk at the start of the ith interval. For example, the number at risk at the start of the interval 24 to 27 days is $n_6 = 5$, making $\hat{q}_6 = 1/5 = 0.2$ and $\hat{P}_6 = 1 - 1/5 = 4/5 = 0.8$, where $n = 10$ and $i = 6$. Thus, the estimated conditional survival probability of surviving beyond 27 days (t_6) given that the AIDS patient was alive on day 24 (t_5) is 0.8. These probabilities are conditional because they apply only to individuals alive at the beginning of the interval (condition = alive at time t_{i-1}). That is, the conditional probability of death $\hat{q}_6$ is

$$\hat{q}_6 = P \text{ (death between 24 and 27 days survived 24 days)}$$
$$= \frac{P \text{ (death between 24 and 27 days)}}{P \text{(survived 24 days)}} = \frac{1/10}{5/10} = \frac{1}{5} = 0.2.$$

The estimated probability of surviving beyond a specific time t_k (denoted $\hat{P}_k$) is the product of the conditional probabilities of surviving each of the

Table 4.2. Product-limit estimated survival probabilities and 95% confidence intervals from the $n = 10$ AIDS survival times (complete data).

i	$t_{i-1} - t_i$	d_i	n_i	$\hat{p}_i$	$\hat{P}_i$	lower	upper
1	0–2	1	10	0.900	0.9	0.473	0.985
2	2–4	1	9	0.889	0.8	0.409	0.946
3	4–14	1	8	0.875	0.7	0.329	0.892
4	14–21	1	7	0.857	0.6	0.252	0.827
5	21–24	1	6	0.833	0.5	0.184	0.753
6	24–27	1	5	0.800	0.4	0.123	0.670
7	27–33	1	4	0.750	0.3	0.071	0.578
8	33–51	1	3	0.667	0.2	0.031	0.475
9	51–60	1	2	0.500	0.1	0.006	0.358
10	60–72	1	1	0.000	0.0	—	—

first k intervals. Thus, the estimated survival probability is

$$\hat{P}_k = \hat{p}_1 \times \hat{p}_2 \times \cdots \times \hat{p}_k = \prod \hat{p}_i = \prod \frac{n-i}{n-i+1}, \qquad i = 1, 2, \ldots, k.$$

The probability $\hat{P}_k$ is called the product-limit estimate or sometimes the Kaplan–Meier estimate of the survival probability. Continuing the AIDS example, the estimated probability of surviving beyond $t_6 = 27$ days ($k = 6$) is

$$\hat{P}_6 = 0.900 \times 0.889 \times 0.875 \times 0.857 \times 0.833 \times 0.800 = 0.4.$$

The product–limit estimated survival probabilities $\hat{P}_1, \hat{P}_2, \ldots, \hat{P}_{10}$ are contained in Table 4.2 based on the survival times of the 10 severely ill AIDS patients.

In fact, the product-limit survival probability calculated from a table of unique and complete survival times is more simply

$$\hat{P}_k = 1 - \frac{k}{n} = \frac{n-k}{n}$$

because

$$\hat{P}_k = \prod \hat{p}_i = \frac{n-1}{n} \times \frac{n-2}{n-1} \times \frac{n-3}{n-2} \times \frac{n-4}{n-3} \times \cdots \times \frac{n-k}{n-k+1}$$

$$= \frac{n-k}{n} = 1 - \frac{k}{n}.$$

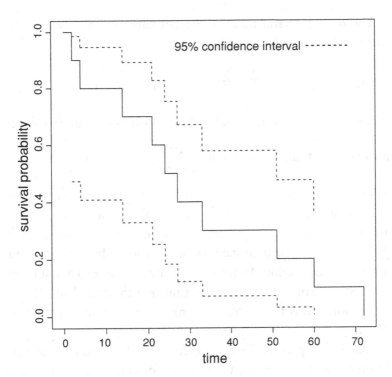

Figure 4.1. Product-limit estimated survival probabilities of extremely ill AIDS patients (complete data) and 95% confidence intervals.

The survival probability $\hat{P}_k$ is naturally estimated by calculating the proportion of individuals who died before or during the kth interval (k/n) and subtracting this value from one ($1 - k/n$). The complementary probability $1 - \hat{P}_k = k/n$ in other statistical contexts is called the *estimated cumulative probability*. The estimate $\hat{P}_k$ from complete data is a typical estimate of a binomial probability. Its estimated variance, for example, is $\hat{v}_k = \hat{P}_k(1 - \hat{P}_k)/n$.

Ten estimated survival probabilities ($\hat{P}_k$-values—Table 4.2) from the AIDS data are displayed in Figure 4.1 along with approximate 95% confidence bounds (to be discussed). The survival "curve" is no more than a series of rectangles (height $= P_{i-1}$ and width $= t_i - t_{i-1}$) placed side by side to display the decreasing pattern of the estimated survival probabilities over time. The survival "curve" is a histogram-like picture of survival probabilities, commonly called a *step function*.

The mean survival time can be estimated in two ways from complete survival data: the usual way,

$$\text{mean survival time} = \bar{t} = \frac{\sum t_i}{n}, \quad i = 1, 2, \ldots, n,$$

or based on the total area enclosed by the estimated survival function,

$$\text{mean survival time} = \text{area} = \hat{\mu} = \sum \hat{P}_{i-1}(t_i - t_{i-1}),$$

where $\hat{P}_0 = 1.0$ and $i = 1, 2, \ldots, n$. The area of the ith rectangle is height $= \hat{P}_{i-1}$ multiplied by width $= t_i - t_{i-1}$, making the area of each rectangles area $= \hat{P}_{i-1}(t_i - t_{i-1})$. The sum of the n rectangles' area $= \sum \text{area}_i = \sum \hat{P}_{i-1}(t_i - t_{i-1})$ is then the total area enclosed by the estimated product-limit survival function. As noted, the total area enclosed by a survival function is an estimate of the mean survival time. For the complete AIDS data, either expression yields the identical estimated mean survival time of $\bar{t} = \hat{\mu} = 30.8$ days.

Clearly, the product-limit estimation of $\hat{P}_k$ and the mean survival time based on the total area are unnecessary for complete survival data, because simple and direct estimates are available. However, these two estimates continue to produce unbiased estimates when the data are not complete (the time of death is not known for all individuals sampled—next topic).

The median time of death is another important summary of survival experience. The estimated median value is that survival time (denoted $\hat{t}_m$) when $\hat{P} = 0.5$. When an estimated survival probability $\hat{P}_i$ does not exactly equal 0.5, an estimate of the median value is the upper bound of the interval containing the survival probability $\hat{P} = 0.5$. The estimated median survival time from the 10 AIDS cases is $\hat{t}_m = 24$ days (Figure 4.2).

Other survival times between the lower and upper bounds of the interval containing the survival probability 0.5 serve equally to estimate the median value. For example, the lower bound or the mean of the upper and lower bounds or a linear interpolation using the endpoints of the median-containing interval are essentially equivalent estimates of the median survival time t_m.

Product-limit estimated survival probabilities are fundamental to analyzing survival data because they are estimated entirely without assumptions about the population that produced the sample of survival times. They are

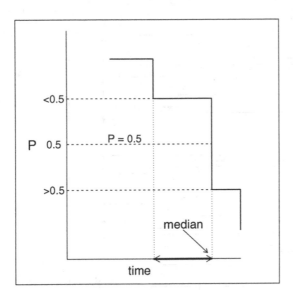

Figure 4.2.   Schematic picture of the estimation of the median survival time.

entirely defined by the properties of the sampled data (called *nonparametric or model-free estimates*).

Censored data

The single most defining characteristic of survival analysis methods is the unbiased description of data when the time of death is not known for all individuals sampled. Figures 4.3 and 4.4 display two versions of the same hypothetical pattern of sampled survival times. A study of survival usually starts by collecting observations over time (labeled "Start" on the figures). The data are not collected on a single day or, in most cases, not even over a short period of time. Rather, as the study progresses, each subject is identified and entered into the study. Subjects are accumulated. The observed individuals either are alive at the end of the study or die at some point during the study.

After a period of time the study ends (labeled "End" on the figures), leaving two kinds of survival times: from the time a subject entered the study to the time of death (called a *complete observation*) or from the time a subject entered the study to the time the study ended (called a *censored observation*). These censored survival times are said to be *right censored*. In studies of survival,

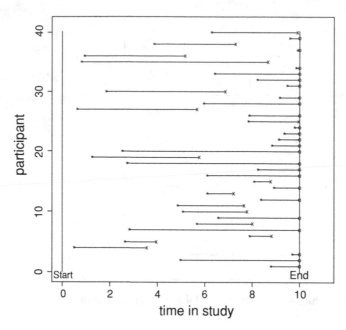

Figure 4.3.   Forty hypothetical survival times as observed over a 10-month study.

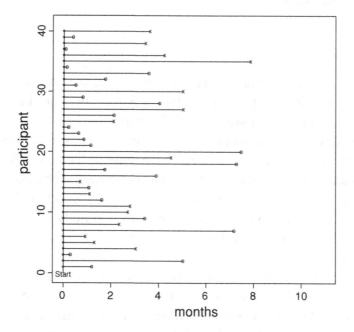

Figure 4.4.   Forty hypothetical survival times measured from entry into the study.

a number of individuals frequently remain alive at the end of the study period resulting in right censored observations. Figure 4.3 depicts 40 hypothetical subjects (repeated from Chapter 2) who entered a study and were observed over a 10-month period. The entry time is represented by an "∗," the time of death by an "x," and the time an observation ends for each individual who did not die by an "o" (censored). The 40 specific survival times (repeated) are

1.2^+	5.0^+	0.3^+	3.0	1.3	0.9	7.2^+	2.3	3.4^+	2.7
2.8	1.6^+	1.1	1.1^+	0.7	3.9^+	1.7^+	7.3^+	4.5	7.5^+
1.2^+	0.9^+	0.6^+	0.2^+	2.1	2.1^+	5.0	4.0^+	0.8^+	5.0
0.5^+	1.8^+	3.6^+	0.1^+	7.9	4.2	0.1^+	3.4	0.4^+	3.6.

The superscript "+" denotes a censored observation (24 censored observations and 16 deaths make up the sample of 40 observed individuals). For example, the first three observations (participants 1, 2, and 3—bottom of Figure 4.3) entered the study 8.8, 5.0, and 9.7 months after the study began. All three participants were alive at the end of the study (censored survival times 1.2^+, 5.0^+, and 0.3^+ months). On the other hand, the fourth, fifth, and sixth subjects (next three lines from the bottom) entered the study 0.3, 2.7, and 8.0 months after the study began and died during the study period (complete survival times 3.0, 1.3 and 0.9 months). Figure 4.4 displays these same 40 individuals but in terms of time from the start of observation until the study participant either was censored ("o") or died ("x").

If this hypothetical study were continued beyond 10 months, the censored individuals would add more survival time and, therefore, estimates made directly from the 40 observed survival times would be biased because the censored individuals would have under-reported survival times. To make the product-limit estimated survival probabilities unbiased, it is necessary to modify the estimation process to account for incomplete survival times. This modification is easily accomplished when the incomplete observations (censored) occur at random, called *noninformative* censoring. That is, the reason that the time of death is not observed is entirely unrelated to the outcome under study. The study simply ended while the observed individual was alive. The statistically important consequence of noninformative censoring is that the underlying probabilities of death are the same for both censored and complete observations. The only relevant difference between the two kinds of observations is that the time of death of the censored individuals is not known. The situation, however, is far more complicated when the reason for censoring is related to the outcome under study. Nonrandom censoring of survival times is an advanced topic and is left to advanced presentations.

From this point on, the term censored refers to right censored observations that are noninformative (random!). However, it is worthwhile to mention several other kinds of incomplete data.

type: Fix time censoring

example: The survival times of a sample of 100 patients are recorded and each person is observed for no more than 10 months (all censored individuals are observed for the same amount of time).

type: Fix number censoring

example: The survival times of a sample of 100 patients are recorded and the study is ended after 75 patients have died.

type: Left censoring

example: Among leukemia patients sampled to study the time from remission to relapse, an observation is left censored when the individual sampled is already in remission. Thus, the "starting time" is not known.

type: Interval censoring

example: Consider the failure times of a sample of patients observed once a year after an experimental treatment. When a failure is observed as part of a yearly visit, the exact time of failure is not known but must have occurred after the last visit and before the present visit.

type: Current status data (a kind of extreme interval censoring)

example: An experimental animal is subjected to a high dose of radiation. After 10 months, the animal is sacrificed and it is determined if a specific tumor is present or absent. The exact failure time ("time-to-tumor") is not known but either it is less than 10 months (tumor present) or more than 10 months (tumor absent).

type: Length-biased survival time

example: Social Security recipients are sampled to determine if individuals with annual incomes greater than $200,000 live longer than average. The mean survival time is likely length-biased because individuals who died before age 65 (not receiving Social Security) will not be included in the sample. Technically, these observations are truncated (missing completely) and not censored.

Analytic techniques for these different kinds of incomplete data are described in detail elsewhere(1).

Another kind of censoring occurs when, for one reason or another, individuals who originally entered into the study are lost from follow-up and their survival status (survived/died) cannot be determined. When the reason for being lost is unrelated to the outcome, the observed survival time is again noninformative right censored. Otherwise, individuals lost to follow-up become a potential source of bias.

When sampled individuals die from causes entirely unrelated to the disease under study, they too can be considered as lost to follow-up. That is, deaths from causes unrelated to the survival outcome (study "endpoint") produce noninformative right censored survival times. In a study of AIDS, for example, a patient who dies in an automobile accident can be treated as a censored observation. Although simple in principle, situations arise in practice where it is difficult to determine whether the cause of death is entirely unrelated to the outcome under study.

When a sample includes censored observations, the estimation process is adjusted to account for the bias incurred from the incomplete survival times. Once this adjustment is made, subsequent analysis proceeds in much the same way as the analysis of most data. For example, the usual statistics (such as means, medians, variances, plots, confidence intervals, and tests) produce useful summaries after the bias incurred from censored data is "removed."

Survival probabilities estimated from censored data

Parallel to the estimation of survival probabilities from complete data (no censored observations), product-limit estimates from data containing censored observations also require that a sequence of intervals be constructed to estimate a sequence of conditional survival probabilities. Again, these intervals are based on the observed time of death (only complete observations). In addition, the survival time of a censored individual is relevant only when the study subject completes the entire interval. Therefore, in each interval only two kinds of individuals enter into the calculation of the conditional survival probability, those who died (creating the endpoint of the interval) and those who survived the entire interval. Thus, an individual whose

Table 4.3. Product-limit estimated survival probabilities for the
$n = 10$ AIDS survival times with three censored observations
(complete $= d = 7$ and censored $= n - d = 10 - 7 = 3$).

i	$t_{i-1} - t_i$	d_i	n_i	$\hat{q}_i$	$\hat{p}_i$	$\hat{P}_i$	$\sqrt{\hat{V}_i}$
1	0–2	1	10	0.100	0.900	0.900	0.095
2	2–4	1	9	0.111	0.889	0.800	0.126
3	4–14	1	8	0.125	0.875	0.700	0.145
4	14–24	1	6	0.167	0.833	0.583	0.161
5	24–27	1	5	0.200	0.800	0.467	0.166
6	27–60	1	2	0.500	0.500	0.233	0.185
7	60–72	1	1	0.000	1.000	0.000	—

censored survival time falls within the interval bounds is not included in
estimates made for that interval or in any subsequent time interval. The
number of persons considered (said to form a *risk set*) is the total number
who survived the entire interval plus those who died (usually one).

The previous AIDS mortality data illustrate where three observations are
censored (21, 33, and 51 days):

survival times (in days):   2, 72, 51^+, 60, 33^+, 27, 14, 24, 4, and 21^+.

As before, the survival time intervals are constructed so that one death occurs
in each interval. The seven risk sets (one for each complete survival time—
Table 4.3) contain 10, 9, 8, 6, 5, 2, and 1 individuals, respectively. When
sampled individuals have identical survival times, they are simply included
in the same risk set.

The interval-specific conditional probability of death ($\hat{q}_i$), the corresponding conditional probability of survival ($\hat{p}_i = 1 - \hat{q}_i$), and the probability of
surviving the first k consecutive intervals ($\hat{P}_k$) are estimated in the same fashion as for complete data. However, the number of individuals in the risk sets
(n_i) does not have a predictable pattern. The number of individuals at risk
depends on the distribution of the randomly occurring censored survival
times. For data containing censored observations, the number of intervals
in the product-limit table is denoted d and is also the number of complete

observations when all survival times are unique. The value d is necessarily less than or equal to the total number of observations (denoted n).

The estimated survival probability $\hat{P}_k$ is again the product of the conditional survival probabilities $\hat{p}_i$. Each conditional estimate $\hat{p}_i$ is unaffected by censored observations (all observations within the interval have the same survival time); therefore, the product of these estimates is also unaffected by the incomplete information from the censored observations. The unbiased product-limit estimate becomes

$$\hat{P}_k = \prod \hat{p}_i = \prod \frac{n_i - d_i}{n_i}, \qquad i = 1, 2, \ldots, k,$$

where d_i represents again the number of deaths among the n_i individuals in the ith risk set (usually one). For unique/complete data, the risk set pattern is $n_i = n - i + 1$ and $d_i = 1$ for all $n = d$ sampled individuals (intervals).

Continuing the AIDS data example, the seven product-limit estimated conditional survival probabilities (one for each observed AIDS death) are given in Table 4.3. For the survival probability P_5 or $P(T \geq 27)$, the product-limit estimate is

$$\hat{P}_5 = \frac{10 - 1}{10} \times \frac{9 - 1}{9} \times \frac{8 - 1}{8} \times \frac{6 - 1}{6} \times \frac{5 - 1}{5}$$
$$= 0.900 \times 0.889 \times 0.875 \times 0.833 \times 0.800 = 0.467.$$

The variance of the distribution of an estimated survival probability is calculated from an expression frequently called *Greenwood's variance* after Major Greenwood, an early biometrician (details are at the end of the chapter). This estimated variance of the distribution of the estimated value $\hat{P}_k$ is given by the expression

$$\hat{V}_k = \text{variance}(\hat{P}_k) = \hat{P}_k^2 \sum \frac{\hat{q}_i}{n_i \hat{p}_i}, \qquad i = 1, 2, \ldots, k,$$

where again n_i represents the number of individuals in the risk set for the ith interval. When $n_i = n - i + 1$ (no censored survival times), the Greenwood variance is the estimated binomial distribution variance. In symbols, the variance of $\hat{P}_k$ becomes $\hat{v}_k = \hat{V}_k = \hat{P}_k(1 - \hat{P}_k)/n$ when $d = n$. Figure 4.5 displays the estimated survival function (step function) accounting for the three censored observations and its 95% confidence intervals.

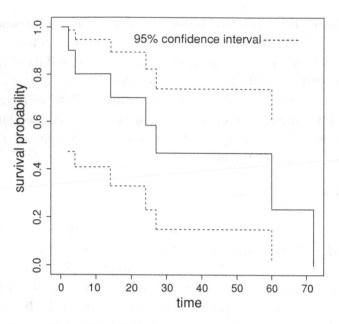

Figure 4.5. Survival probabilities (in days) of extremely ill AIDS patients (censored data).

A bit of care is necessary to construct accurate confidence interval bounds from a product-limit estimated survival probability and its estimated variance. The expression for a 95% confidence interval based directly on an estimated parameter and the normal distribution is estimate $\pm$ $1.960\sqrt{\text{variance (estimate)}}$. However, the accuracy of a confidence interval is frequently improved by calculating the confidence bounds from a function of a parameter and using these limits to derive the confidence interval bounds for the parameter (Chapter 1). The confidence interval for the survival probability P_k is such a case. The function is $\hat{s}_k = \log[-\log(\hat{P}_k)]$. The estimate $\hat{s}_k$ has a more normal-like distribution than the distribution of $\hat{P}_k$. The estimated variance of this approximate normal distribution is

$$\text{variance}(\hat{s}_k) = \text{variance}(\log[-\log(\hat{P}_k)])$$

$$= \frac{\hat{V}_k}{[\hat{P}_k \log(\hat{P}_k)]^2} \quad (\hat{V}_k = \text{the Greenwood variance}).$$

This expression and many of the variances that follow are derived with statistical/mathematical tools described in Chapter 3. The calculation of the approximate 95% confidence bounds based on an estimated survival

Table 4.4. Approximate 95% confidence intervals for the six estimated survival probabilities $\hat{P}_k$ for the example AIDS data ($n = 10$–Table 4.3 and Figure 4.5).

k	Lower	$\hat{P}_k$	Upper
1	0.473	0.900	0.985
2	0.409	0.800	0.946
3	0.329	0.700	0.892
4	0.230	0.583	0.821
5	0.150	0.467	0.737
6	0.015	0.233	0.606

probability $\hat{P}_k$) starts with the usual normal-based confidence interval but constructed from the transformed survival probability ($\hat{s}_k$), where

$$A_k = \text{lower bound} = \hat{s}_k - 1.960\sqrt{\text{variance}(\hat{s}_k)}$$

and

$$B_k = \text{upper bound} = \hat{s}_k + 1.960\sqrt{\text{variance}(\hat{s}_k)},$$

making A_k and B_k the 95% confidence interval bounds for the transformed parameter s_k from the kth interval. Then, the bounds A_k and B_k become the basis for estimating the approximate confidence interval bounds for the survival probability P_k. Specifically, they are

$$\text{lower bound} = e^{-\exp(A_k)} \quad \text{and} \quad \text{upper bound} = e^{-\exp(B_k)}.$$

As required, the estimated survival probability is $\hat{P}_k = e^{-\exp(\hat{s}_k)}$. The notation $\exp(x)$ is used in place of the term e^x simply to make the expression easier to read. Six examples of approximate 95% confidence intervals based on the estimated AIDS survival probabilities (Table 4.3) are given in Table 4.4.

To illustrate the valuable role a transformation plays in construction of an accurate confidence interval, an example (based on $n = 100$ hypothetical survival times) is displayed in Figure 4.6. The first two plots (first row)

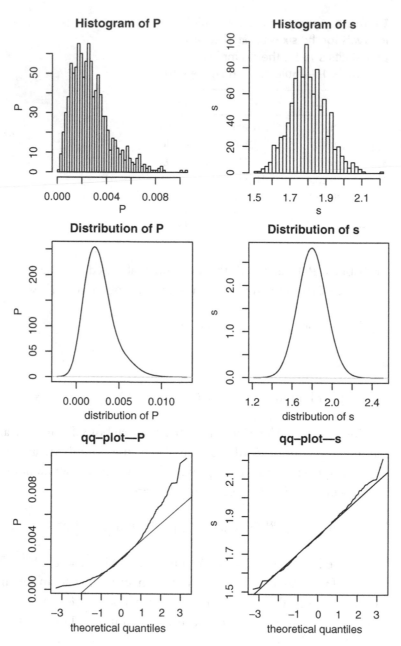

Figure 4.6.   A comparison of the distributions of the survival probabilities $\hat{P}_k$ and their transformed values $\hat{s}_k = \log(-\log[\hat{P}_k])$ for $n = 100$ hypothetical survival times.

are histograms of the distributions of the survival probabilities $\hat{P}_k$ and the transformed values $\hat{s}_k = (\log - \log[\hat{P}_k])$. The second row displays the same distributions smoothed to emphasize more clearly the asymmetry ($\hat{P}_k$) and symmetry ($\hat{s}_k$) of these two distributions. The third row contains specialized plots, called *quantile plots*, to compare the distributions of values of $\hat{P}_k$ and $\hat{s}_k$ to a normal distribution (the straight line indicates a perfect normal distribution).

An additional feature of this log-log transformation of $\hat{P}_k(\hat{s}_k)$ is that the lower confidence interval bound is never less than 0.0 and the upper bound is never greater than 1.0, which is consistent with a survival probability that is also always between these two bounds.

Notice that as the estimated survival probabilities decrease, the lengths of the confidence intervals generally increase. This increase is primarily due to the reduction in the number of observations associated with the estimation of the smaller survival probabilities (longer survival times).

An approximate normal-based confidence interval generated by each estimate $\hat{P}_k$ is formally called a *pointwise confidence interval*. At each survival time (t_k), the probability is approximately 0.95 that the underlying survival probability P_k is contained in the interval constructed from its estimate $\hat{P}_k$. Connecting a sequence of lower and upper bounds surrounding the estimated survival function does not form a confidence band (Figures 4.1 and 4.5).

A confidence band is constructed so that the probability is 0.95 that the lower and upper confidence interval bounds contain the entire survival curve. Pointwise confidence intervals each contain a specific survival probability P_k with probability 0.95. The probability that a series of such intervals simultaneously contains all values of the survival probabilities P_k is necessarily less than 0.95. Thus, the width of a 95% confidence band is larger than the limits created by connecting a series of pointwise 95% confidence interval bounds. The theory and a table of specialized values necessary to construct this larger confidence band exist elsewhere [2].

The process used to estimate the mean survival time (denoted μ) from complete data applies equally to data containing censored observations. The area under the estimated survival curve remains an estimate of the mean value. When the survival probabilities are estimated so that they are not

Table 4.5. Components of the mean survival time estimated from the product-limit table of the AIDS data (Table 4.3).

Interval	$i = 1$	$i = 2$	$i = 3$	$i = 4$	$i = 4$	$i = 6$	$i = 7$
Heights: $\hat{P}_{i-1}$	1.00	0.900	0.800	0.700	0.583	0.467	0.233
Widths: $t_i - t_{i-1}$	2	2	10	10	3	33	12
Areas: $\hat{P}_{i-1}(t_i - t_{i-1})$	2.00	1.80	8.00	7.00	1.75	15.40	2.80

biased by the influences of censored data, the survival curve constructed from these probabilities is also not biased by the presence of censored data. Using the AIDS data illustrates (Table 4.5). The estimated mean survival time is

$$\text{mean survival time} = \hat{\mu} = \text{area} = \sum \text{area}_i = \sum \hat{P}_{i-1}(t_i - t_{i-1})$$
$$= 1.000(2) + 0.900(2) + \cdots + 0.467(33) + 0.233(12)$$
$$= 2.00 + 1.80 + \cdots + 15.40 + 2.80 = 38.750 \text{ days},$$

where the estimated survival probabilities $\hat{P}_i$ are given in Table 4.3 and displayed in Figure 4.5.

The estimated variance of the distribution of the estimated mean (area under the survival curve) requires a sum of a series of values denoted A_k. These values are related to the d rectangular areas (area$_i$) that make up the product-limit estimated survival curve and are defined by

$$A_k = \sum \text{area}_i = \sum \hat{P}_{i-1}(t_i - t_{i-1}),$$
$$i = k+1, k+2, \ldots, d \quad (\text{note}: A_0 = \hat{\mu}).$$

The estimated variance of the distribution of the estimated mean survival time $\hat{\mu}$ becomes

$$\text{variance}(\hat{\mu}) = \sum \frac{A_k^2}{n_k(n_k - 1)}, \quad k = 1, 2, \ldots, d,$$

where again d represents the number of intervals in the product-limit table (the number of complete observations if no identical survival times occur). For the example data from the $n = 10$ AIDS patients (Table 4.3), the estimated variance is

$$\text{variance}(\hat{\mu}) = 15.006 + 16.965 + \cdots + 3.920 = 78.690.$$

The details of the computation are given in Table 4.6.

Table 4.6. Details of the calculation of the variance of the product-limit estimated mean value $\hat{\mu} = 38.750$ days.

k	$i=2$	$i=3$	$i=4$	$i=5$	$i=6$	$i=7$	A_k	m_k^*	A_k^2/m_k
1	1.80	8.00	7.00	1.75	15.40	2.80	36.75	10×9	15.006
2	—	8.00	7.00	1.75	15.40	2.80	34.95	9×8	16.965
3	—	—	7.00	1.75	15.40	2.80	26.95	8×7	12.970
4	—	—	—	1.75	15.40	2.80	19.95	6×5	13.267
5	—	—	—	—	15.40	2.80	18.20	5×4	16.562
6	—	—	—	—	—	2.80	2.80	2×1	3.920

* $m_k = n_k(n_k - 1)$, where n_k is the number of at-risk individuals in the kth row of the product-limit table.

The usual normal-based approximate 95% confidence interval is $\hat{\mu} \pm 1.960\sqrt{\text{variance}(\hat{\mu})} = 38.750 \pm 1.960\sqrt{78.690}$ or $(21.363, 56.137)$. When the data are complete and unique ($n = d$ and no identical survival times occur), the variance of the estimated mean value is given by variance$(\hat{\mu}) = \sum(\hat{t}_i - \hat{t})^2/n^2$, indicating a slight bias. To correct for this bias, the estimated variance is multiplied by $n/(n-1)$.

A mean value estimated to summarize survival experience has two disadvantages. It is not defined when the longest survival time is censored (area $= ?$). Of more importance, the distribution of survival times is typically asymmetric (skewed to the right—long right "tail"), making a mean survival time less representative of a "typical" observation than a median value. The median value is a more representative summary when the sampled distribution is not symmetric.

The median survival time estimated from censored survival data follows the same pattern described for complete data but is based on a product-limit table that accounts for incomplete observations. The interval that contains the survival probability $P = 0.5$ is located and the estimated median survival time is again the upper bound of that interval. Two additional interval bounds are also useful, the next smallest bound (denoted $\hat{t}_l$) and the next largest bound (denoted t_u). For the AIDS data, the estimated median survival time $\hat{t}_m$ is 27, making $\hat{t}_l = 24$ and $t_u = 60$ (Table 4.3). An estimated variance of the distribution of the estimated median survival survival time is constructed from these three values. It is

$$\text{variance}(\hat{t}_m) = \left[\frac{t_1 - t_u}{\hat{P}_1 - \hat{P}_u}\right]^2 \hat{V}_m,$$

where $\hat{P}_1$ and $\hat{P}_u$ are the estimated survival probabilities corresponding to the survival times t_1 and t_u. As before, the quantity denoted $\hat{V}_m$ represents the Greenwood estimated variance of the estimated value $\hat{P}_m$, where m denotes the interval containing the survival probability of 0.5.

For the censored AIDS data, the estimated median survival time $\hat{t}_m$ is 27 with estimated variance

$$\text{variance}(\hat{t}_m) = \left[\frac{24 - 60}{0.583 - 0.233}\right]^2 0.0275 = 290.743.$$

As noted, the values used to calculate the variance of a median value are found in the product-limit table. For the example, from Table 4.3 (row = $m = 5$), $\hat{P}_4 = 0.583$, $\hat{P}_5 = 0.467$, $\hat{P}_6 = 0.233$, $\hat{t}_4 = 24$, $\hat{t}_5 = 27$, $\hat{t}_6 = 60$, and $\hat{V}_5 = (0.166)^2 = 0.0275$.

The approximate 95% confidence interval $\hat{t}_m \pm 1.960 \sqrt{\text{variance}(\hat{t}_m)}$ based on the estimated median value $\hat{t}_5 = 27$ is $(-6.420, 60.420)$. When the normal approximation yields an impossible negative bound, it is some-times conventionally replaced with a zero. The approximate confidence inter-val for the median survival time is then reported as $(0.0, 60.420)$. Of more fundamental importance, a negative bound serves as a warning that the accu-racy of the normal distribution as an approximation is failing. The negative bound and the wide confidence interval (lack of precision) for the AIDS data are not surprising, because the estimated median survival time is based on only 10 observations with three censored survival times (seven complete observations—$d = 7$). More effective estimates of the median survival time and its variance will be discussed (Chapter 5 and beyond).

Estimates exist of the hazard function from product-limit tables [3]. How-ever, such estimated hazard functions are subject to considerable sampling variation because they are estimated from intervals that typically contain only a few observations (usually only one death). In addition, the interpretation of the resulting estimated hazard function is not simple. Like the median survival time, extremely efficient estimates of the hazard function will be discussed that take advantage of the entire data set (Chapter 5 and beyond).

Although a hazard function is not easily interpreted or accurately estimated from product-limit estimated probabilities, a related summary, called
the *cumulative hazard function*, is an alternative way to describe survival data
or to contrast survival experiences among different groups. Parallel to the
product-limit estimated survival probability, the cumulative hazard function
is a summary derived from cumulative estimates based again on the interval
specific estimated conditional survival probabilities ($\hat{p}_i$). This summary provides another view of the survival pattern but its role in the analysis of
survival data is not fundamentally different from that of a product-limit
estimated survival curve. The estimated cumulative hazard function (sometimes referred to the *cumulative disease incidence or mortality function*) is
defined by the expression

$$\hat{H}(t_k) = -\log(\hat{P}_k) = -\log\left(\prod \hat{p}_i\right) = -\sum \log(\hat{p}_i), \qquad i = 1, 2, \ldots, k,$$

where, as before, $\hat{P}_k$ is the product-limit estimated survival probability for the
kth interval constructed from the estimated conditional survival probabilities
$\hat{p}_i$. Because $\log(p) = \log(1 - q) \approx -q$, an approximate cumulative hazard
function is more simply expressed as

$$\hat{H}^*(t_k) = \sum \hat{q}_i, \qquad i = 1, 2, \ldots, k.$$

For most human survival data, $\hat{H}(t_k) \approx \hat{H}^*(t_k)$. The interval-specific
approximate values of $\hat{H}^*(t_k)$ are

$$\hat{H}^*(t_1) = \hat{q}_1$$
$$\hat{H}^*(t_2) = \hat{q}_1 + \hat{q}_2$$
$$\hat{H}^*(t_3) = \hat{q}_1 + \hat{q}_2 + \hat{q}_3$$
$$-\!-$$
$$-\!-$$
$$\hat{H}^*(t_k) = \hat{q}_1 + \hat{q}_2 + \cdots + \hat{q}_k,$$

indicating the reason that $\hat{H}(t_k)$ is called the estimated cumulative hazard
function ($h(t) \approx q$—Chapter 1). For the 10 AIDS patients (Table 4.3), the
estimated values $\hat{H}(t_k)$ and $\hat{H}^*(t_k)$ are presented in Table 4.7 and Figure 4.7.

Table 4.7. Estimated cumulative hazard function $\hat{H}(t_i)$ and $\hat{H}^*(t_i)$ for the 10 AIDS survival times (Table 4.3).

i	$t_{i-1} - t_i$	d_i	n_i	$\hat{P}_i$	$\hat{H}(t_i)$	$\hat{H}^*(t_i)$
1	0–2	1	10	0.900	0.105	0.100
2	2–4	1	9	0.889	0.223	0.211
3	4–14	1	8	0.875	0.357	0.336
4	14–24	1	6	0.833	0.539	0.503
5	24–27	1	5	0.800	0.762	0.703
6	27–60	1	2	0.500	1.455	1.203
7	60–72	1	1	0.000	—	2.203

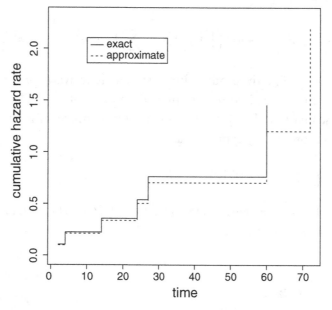

Figure 4.7.   Estimated cumulative hazard function $\hat{H}(t_k)$ and $\hat{H}^*(t_k)$ for the 10 AIDS survival times (Table 4.3).

APPENDIX

Statistical tools: Greenwood's variance expression

A not exactly straightforward application of the expression for the variance of the logarithm of a variable (described in Chapter 3) produces an estimate for the variance of an estimated survival probability (Greenwood's variance expression).

Step 1. When $\hat{p}_i$ represents an estimated probability from a binomial distribution, then

$$\text{variance}[\log(\hat{p}_i)] \approx \left[\frac{1}{\hat{p}_i}\right]^2 \text{variance}(\hat{p}_i) = \left[\frac{1}{\hat{p}_i}\right]^2 \frac{\hat{p}_i(1-\hat{p}_i)}{n_i} = \frac{\hat{q}_i}{n_i\hat{p}_i},$$

where $\hat{q}_i = 1 - \hat{p}_i$ and the variance of the distribution of $\hat{p}_i$ is estimated by $\hat{p}_i(1-\hat{p}_i)/n_i$ (Chapter 1).

Step 2. For the product-limit estimated survival probability $(\hat{P}_k)$, the estimated variance is

$$\text{variance}(\hat{P}_k) \approx \hat{P}_k^2 \text{ variance}[\log(\hat{P}_k)].$$

Step 3. The logarithm of the product-limit estimate of $\hat{P}_k$ is a sum of the logarithms of k specific conditional survival probabilities $\hat{p}_1, \hat{p}_2, \hat{p}_3, \ldots, \hat{p}_k$ and is

$$\log(\hat{P}_k) = \log\left(\prod \hat{p}_i\right) = \sum \log(\hat{p}_i), \qquad i = 1, 2, \ldots, k.$$

Step 4. Putting steps 1, 2, and 3 together gives Greenwood's expression,

$$\text{variance}(\hat{P}_k) \approx \hat{P}_k^2 \text{ variance}[\log(\hat{P}_k)] = \hat{P}_k^2 \text{ variance}\left[\sum \log(\hat{p}_i)\right]$$

$$= \hat{P}_k^2 \sum \text{variance}[\log(\hat{p}_i)] \approx \hat{P}_k^2 \sum \frac{\hat{q}_i}{n_i\hat{p}_i}, \qquad i = 1, 2, \cdots, k.$$

Exponential survival time probability distribution

A product-limit estimate of a survival probability is model-free. Thus, the resulting estimates do not depend on assumptions or require knowledge about the population that produced the sampled survival data. The value estimated is entirely determined by the data. A totally different approach requires that a specific parametric model describe the sampled population. The *exponential survival time probability distribution* is one such model. It is a simple but theoretical distribution that completely defines a survival probability based on a single parameter (denoted λ). Specifically, this survival function is

survival probability $= P(T \geq t) = S(t) = e^{-\lambda t}$.

For example, for time $t = 20$ years, the survival probability is $P(T \geq 20) = S(20) = e^{-0.04(20)} = 0.449$ when the exponential distribution parameter is $\lambda = 0.04$ (Chapter 1, Figure 1.1). Thus, for a randomly sampled individual whose survival time is described by these exponential survival time data ($\lambda = 0.04$), the probability of surviving beyond time $t = 20$ years is 0.449. For any other value of t, the corresponding survival probability is similarly calculated. The survival time distribution depends entirely on a single parameter (in the example, $\lambda = 0.04$) to completely describe the population sampled. Figure 5.1 displays the geometry of three exponential survival functions ($\lambda = 0.04, 0.15$, and 1.0).

The exponential function $f(x) = e^{-\lambda x}$ is found in a variety of contexts other than survival analysis. For example, it is used to describe the physics of heat loss from an object (Newton's law of cooling), the loss of electrical charge, the oscillations of a spring, and the interest accumulated in a bank account. In general terms, this relationship arises when a rate of change is proportional only to the size of the quantity that is changing.

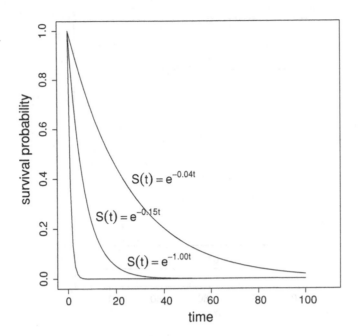

Figure 5.1.   Examples of three exponential survival distributions ($\lambda = 0.04, 0.15$, and 1.0).

Survival times sampled from an exponential distribution have a constant hazard rate, and for a constant hazard rate, the survival times have an exponential distribution. The constant hazard rate associated with exponentially distributed survival times is

$$\text{hazard function} = h(t) = \lambda.$$

Remarkably, the parameter that defines the survival function $S(t)$ is also the hazard rate $h(t)$, namely λ. Figure 5.2 displays the three clearly constant hazard functions corresponding to the three exponential survival distributions in Figure 5.1.

The property that the exponential hazard function is constant follows directly from the general definition of a hazard rate. Specifically,

$$h(t) = -\frac{\frac{d}{dt} S(t)}{S(t)} = -\frac{\frac{d}{dt} e^{-\lambda t}}{e^{-\lambda t}} = \frac{\lambda e^{-\lambda t}}{e^{-\lambda t}} = \lambda.$$

Application of the definition of an instantaneous relative rate (Chapter 1) shows that the hazard rate is constant (the same value regardless of the value of t). This fact is mathematically unequivocal but not particularly intuitive.

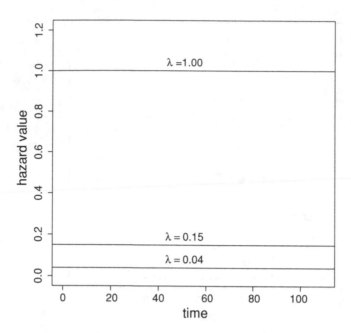

Figure 5.2. Examples of hazard rates from three exponential distributions ($\lambda = 0.04, 0.15$, and 1.0).

Two justifications of the relationship between $S(t)$ and $h(t)$ follow, based on less abstract arguments.

Justification 1

The value of the expression $(1 + 1/n)^n$ is approximately equal to 2.714 for moderately large values of n (n in the neighborhood of 200). When n becomes infinite, the value of this expression is denoted by the symbol e ("e" for exponential and in honor of the major contributions of the mathematician Leonhard Euler). The value of e has been calculated to many nonrepeating decimal places as $2.71828182846\cdots$. Furthermore, the more general expression

$$\left(1 - \frac{x}{n}\right)^n \approx e^{-x}$$

is also approximate for large n and exact for infinitely large n.

Suppose a time interval $(0, t)$ is divided into n equal subintervals of length δ ($n\delta = t$) and the conditional probability of surviving each subinterval is exactly the same (represented previously by p_i and for this special case $p_i = p$

for all n subintervals). Then the survival probability is approximately

$$S(t) = P(T \geq t) = \prod_{i=1}^{n} p_i = p^n = (1-q)^n$$

$$= \left(1 - \frac{nq}{n}\right)^n \approx e^{-qn} = e^{-\lambda\delta n} = e^{-\lambda t}$$

because the hazard rate $\lambda = q/\delta$ is constant ($q = 1 - p$ is constant) for all subintervals and $\delta n = t$. The survival probability $S(t)$ becomes exactly $e^{-\lambda t}$ when n is infinitely large, which is another way of saying that the exponential survival distribution becomes $S(t) = e^{-\lambda t}$ when time t is a continuous variable ($\delta = 0$).

Justification 2

The probability of surviving from time $= 0$ to time $= t_2$ divides into two obvious intervals, 0 to t_1 and t_1 to t_2. Thus, the probability of surviving beyond time t_2 is

$$S(t_2) = P(T \geq t_2) = P(T \geq t_2 \mid T \geq t_1)P(T \geq t_1).$$

For exponentially distributed survival times, this survival probability becomes

$$S(t_2) = e^{-\lambda t_2} = P(T \geq t_2 \mid T \geq t_1)e^{-\lambda t_1},$$

and for the interval t_1 to t_2, the conditional survival probability is

$$P(T \geq t_2 \mid T \geq t_1) = e^{-\lambda(t_2-t_1)}.$$

This straightforward relationship indicates the key property of exponentially distributed survival times. The probability of surviving from time t_1 to t_2 depends only on the magnitude of the difference $t_2 - t_1$ and not the actual values of t_1 and t_2. For example, when $t_2 - t_1 = 20$ years ($\lambda = 0.04$), the survival probability $P(T \geq t_2 \mid T \geq t_1) = e^{-0.04(20)} = 0.449$ for $t_2 = 100$ and $t_1 = 80$ or for $t_2 = 25$ and $t_1 = 5$. Therefore, the probability of surviving 20 additional years is the same for 80-year-olds as for 5-year-olds. The probability is 0.449 for any 20-year period. In general, exponential survival describes the risk for individuals who do not age or objects that do not wear out. In other words, the hazard rate is constant.

This "no memory" property obviously does not apply across the full spectrum of the human lifetime. Clearly, the risk of death differs considerably between 80-year-olds and 5-year-olds (Chapter 2, Figure 2.2). However, some kinds of accidents appear to have constant risk (constant hazard rate). Computer chips do not wear out and, therefore, their failure times are accurately described by an exponential distribution. Occasionally, special populations experience essentially constant hazard rates. For example, an approximately constant hazard rate frequently describes the mortality risk of extremely ill or extremely old individuals. Most relevant to epidemiologic and medical data, over short time intervals, the changes in human mortality rates are typically small, making risk approximately constant (hazard rate $= \lambda$).

A classic situation described by an exponential probability model (constant hazard rate model) is the survival of a wine glass. The probability that a wine glass is broken ("failure") does not depend on how long the wine glass has lasted or how many times it has been used. Wine glasses do not wear out. Glasses are broken with the same probability (at random) regardless of age or amount of use. Unlike most things, wine glasses have the same risk of being broken whether they are new or old. Thus, the lifetime of a wine glass is accurately described by a constant hazard rate for all glasses at all times. An exponential distribution, therefore, accurately predicts the probability that a wine glass will last a specified amount of time. The lifetime of a wine glass (in months) might look like this:

Present age	Additional lifetime
0	40
20	40
40	40
60	40
80	40

A unique and defining property of survival times with an exponential distribution (a constant hazard rate) is that all individuals at a specific time t_0 have the same mean additional survival time. When the hazard rate is constant, individuals who have survived 80 weeks or 40 weeks or 5 weeks have the same mean additional survival time (denoted μ). More technically, the survival probabilities at any time t_0 continue to be exponentially distributed with the same hazard rate (λ), namely $S(t) = e^{-\lambda(t-t_0)}$. Regardless of the value of $t - t_0$, the hazard rate remains λ (constant). Risk necessarily remains, therefore, the same regardless of past survival experience ("no memory"), putting all individuals on an even footing in terms of the amount of future survival time (such as wine glasses).

The previously analyzed 10 AIDS survival times (denoted again t_i, where $n = 10$, $d = 7$ complete, and $n - d = 3$ censored observations) are as follows:

survival times (in days):   2, 72, 51^+, 60, 33^+, 27, 14, 24, 4, and 21^+.

As before, unbiased estimates based on these data must account for the incomplete survival times. When these AIDS survival times are sampled from a population that is at least approximately described by an exponential survival distribution, accounting for the influence of the incomplete survival times is relatively simple.

To begin, the three incomplete values must be randomly censored (noninformative). These censored values are then made "complete." The survival times become $51 + \mu$, $33 + \mu$, and $21 + \mu$, where μ is the mean survival time of the underlying exponential survival distribution. Because the hazard rate is constant and censoring noninformative, all surviving individuals have identical remaining mean survival time μ, regardless of the previously observed survival time t. The estimated mean survival time, calculated from the seven complete and three "completed" survival times, is then

$$\hat{\mu} = \frac{2 + 72 + (51 + \hat{\mu}) + 60 + (33 + \hat{\mu}) + 27 + 14 + 24 + 4 + (21 + \hat{\mu})}{10}$$

or

$$\hat{\mu} = \frac{308 + 3\hat{\mu}}{10}.$$

Solving for $\hat{\mu}$ gives an unbiased estimated mean survival time of $\hat{\mu} = 308/7 = 44$ days. When the incomplete nature of the censored survival times is ignored, the biased estimate (likely too small) is $\bar{t} = \sum t_i/n = 308/10 = 30.8$ days.

From another point of view, the directly estimated mean value $\bar{t} = \sum t_i/n$ is likely too small because the sum of all observed survival times ($\sum t_i$) includes censored individuals who would have added more survival time if the observation period had been extended. Therefore, including an estimate of the "missing" survival time gives

$$\hat{\mu} = \frac{\sum t_i + (n - d)\hat{\mu}}{n}, \quad i = 1, 2, \ldots, n = \text{total number of observations,}$$

where $(n - d)\hat{\mu}$ estimates the total amount of unobserved survival time accumulated by the censored individuals. Each of the $n - d$ censored individuals is unobserved for, on the average, μ days. The sum $\sum t_i$ is the total amount of observed survival times accumulated by both censored and uncensored individuals. Adding the estimated "missing" survival time to the numerator of the estimate of the mean survival time produces an unbiased estimate of the total survival time (observed time + estimated "missing" time). The estimate of the underlying mean value μ is unbiased as long as the hazard rate of the sampled population is constant and the exponential distribution sampled is the same for censored and complete observations (noninformative censoring). Consequently, the mean survival time is the same for both censored and complete observations. Solving the previous expression for $\hat{\mu}$ gives the estimated mean survival time as

$$\hat{\mu} = \frac{\sum t_i}{d}, \qquad i = 1, 2, \ldots, n = \text{number of observations},$$

where the total observed survival time (complete plus censored values) is divided by the number of complete observations d and not the total number of observations n. Dividing by the smaller value d increases the estimated mean value, compensating for the unobserved survival time. The mean survival time of the 10 AIDS patients is again $\hat{\mu} = 308/7 = 44$ days.

An estimated variance of the estimated mean survival time $\hat{\mu}$ is

$$\text{variance}(\hat{\mu}) = \frac{\hat{\mu}^2}{d}$$

and provides an estimate of the variance of the distribution of $\hat{\mu}$ estimated from either complete ($d = n$) or censored ($d < n$) exponential survival data.

Notice that the precision of the estimated survival time is not affected by the number of censored observations ($n - d$). Thus, accounting for the influence of the censored observations is an issue of bias and not precision. Incidentally, the estimate $\hat{\mu}$ is the maximum likelihood estimate of the mean survival time μ. That is, maximizing the likelihood function produces the same estimated mean value and variance.

As noted previously (Chapter 4), the accuracy of an approximate confidence interval can be improved by constructing the interval bounds from a transformed value of the estimate. For the estimate $\hat{\mu}$, the transformed value is $\log(\hat{\mu})$. In general, the accuracy of the coverage probability of an approximate confidence interval based on a normal distribution improves when the variance of the estimate does not depend on the estimated value (estimated value and its variance are unrelated). The logarithm of the estimated mean survival time $\log(\hat{\mu})$ has this property. The estimated variance of the distribution of the logarithm of estimated mean survival time is

$$\text{variance}(\log[\hat{\mu}]) = \frac{1}{\hat{\mu}^2} \text{variance}(\hat{\mu}) = \frac{1}{\hat{\mu}^2} \frac{\hat{\mu}^2}{d} = \frac{1}{d} \quad \text{(Chapter 3)}$$

and depends only on the number of complete observations (d), making it independent of the estimate $\hat{\mu}$. Thus, normal-distribution-based confidence interval bounds from exponentially distributed survival times using the transformed estimate $\log(\hat{\mu})$ are

$$A = \text{lower bound} = \log(\hat{\mu}) - 1.960\sqrt{\frac{1}{d}}$$

and

$$B = \text{upper bound} = \log(\hat{\mu}) + 1.960\sqrt{\frac{1}{d}}$$

and lead to approximately 95% confidence interval bounds for the underlying mean value μ given by (e^A, e^B), calculated from the estimated mean value $\hat{\mu}$. Another feature of confidence bounds based on a logarithmic transformation is that the lower bound is never negative, which is important because the mean survival time is also never negative.

The San Francisco Men's Health Study (SFMHS) was established in 1983 to conduct a population-based study of the epidemiology and natural history of the newly emerging disease Acquired Immunodeficiency Syndrome (AIDS). The study design was a prospective investigation based on a random sample of single men 25–54 years of age living in a highly affected San Francisco neighborhood known as the "Castro District." The group studied ultimately contained 1,034 men, which resulted from a roughly 70% response rate from a random sample of 19 selected San Francisco census tracts. Study participants were examined and questioned every six months from July 1984 through June 1994. The data explored here are a small subset consisting of the individuals who entered the study during the first year (see Table A.2, p. 248). Survival time is defined as the number of months from diagnosis of AIDS to death for these data.

For the 174 white men from this SFMHS cohort ($d = 155$ deaths and $n - d = 19$ censored individuals), the estimated mean survival time is $\hat{\mu} = \sum t_i / d = 3578/155 = 23.084$ months, based on the assumption that survival times were sampled from an exponential probability distribution. The approximate 95% confidence interval bounds using the transformed estimate $\log(\hat{\mu}) = \log(23.084) = 3.139$ are

$$A = \text{lower bound} = 3.139 - 1.960\sqrt{\frac{1}{155}} = 2.982$$

and

$$B = \text{upper bound} = 3.139 + 1.960\sqrt{\frac{1}{155}} = 3.297.$$

The approximately 95% confidence interval bounds for the mean survival time μ become $(e^A, e^B) = (e^{2.982}, e^{3.297}) = (19.721, 27.020)$.

Higher risk clearly dictates lower survival time and vice versa, as already noted (Chapters 1, 2, and 4). Expressed formally for the exponential survival distribution, the relationship between a mean survival time and a hazard rate is

$$\text{hazard rate} = \lambda = \frac{1}{\mu} \quad \text{or} \quad \text{mean survival time} = \mu = \frac{1}{\lambda}.$$

Again risk is inversely related to survival time. More technically,

$$\mu = \text{mean survival time} = \text{area} = \int_0^\infty S(u)du = \int_0^\infty e^{-\lambda u}du = \frac{1}{\lambda}$$

because the area enclosed by the survival function $S(t)$ is the mean survival time (Chapter 1). The geometry of this reciprocal relationship between risk and survival is displayed in Figure 5.3 for three survival functions with hazard rates ($\lambda = 0.04, 0.1$, and 0.5). As required, the larger hazard rates (risk) produce survival curves with smaller enclosed areas (mean survival times) and vice versa.

A natural estimate of a constant hazard rate follows as the number of deaths divided by the total person-time, or

$$\hat{\lambda} = \frac{1}{\hat{\mu}} = \frac{d}{\sum t_i}.$$

For the SFMHS data, the estimated constant hazard rate is $\hat{\lambda} = 1/23.084 = 155/3578 = 0.043$. Because $\hat{\mu}$ is a maximum likelihood estimate, the

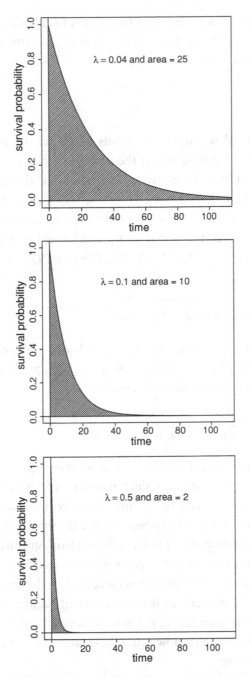

Figure 5.3.   Three examples of exponential survival functions ($\lambda = 0.04, 0.1$, and 0.5) displaying mean survival time (shaded area).

estimated rate $\hat{\lambda}$ (a function of $\hat{\mu}$) is also a maximum likelihood estimate. The maximum likelihood estimated variance of the approximate normal distribution of the estimate $\hat{\lambda}$ is

$$\text{variance}(\hat{\lambda}) = \frac{\hat{\lambda}^2}{d}.$$

Because $\hat{\lambda} = 1/\hat{\mu}$, confidence interval bounds based on the estimated rate $\hat{\lambda}$ are the same reciprocal function of the previously calculated lower and upper bounds derived from the estimated mean survival time $\hat{\mu}$.

Specifically, the *lower bound* $= 1/e^B = 1/e^{3.297} = 1/27.020 = 0.037$ and *upper bound* $= 1/e^A = 1/e^{2.982} = 1/19.721 = 0.051$ produce the approximate 95% confidence interval (0.037, 0.051) for the hazard rate based on the estimated rate $\hat{\lambda} = 1/\hat{\mu} = 1/23.084 = 0.043$. The same 95% confidence interval is

$$\exp\left\{\log(\hat{\lambda}) \pm 1.960\sqrt{\text{variance}(\log[\hat{\lambda}])}\right\} = \hat{\lambda}e^{\pm 1.960/\sqrt{d}}.$$

The exponential distribution of survival times is a theoretical justification of the widely used estimate of an average mortality rate (Chapter 1). That is, an average mortality rate is commonly estimated by

$$\text{average rate} = \frac{\text{total deaths}}{\text{total person-time}} = \frac{d}{\sum t_i} = \hat{\lambda} = \frac{1}{\hat{\mu}},$$

where $\sum t_i$ is the total observed person-time (again censored plus complete survival times). This estimate accurately reflects mortality risk only when the underlying hazard rate is constant. A single estimate requires that the quantity estimated be a single value. To create data where the rate is approximately constant, mortality and disease data are typically collected from groups that are as homogeneous as possible (for example, age-, sex-, race-specific groups). A homogeneous group of sampled individuals is more likely to have at least an approximately constant risk of death or disease (constant hazard), reflected accurately by the single estimated rate $\hat{\lambda}$, particularly when the time interval considered is small. Experience has shown that these estimates are generally effective in human populations.

Under the assumption that the data are sampled from a population with a constant hazard rate, an estimate of the median survival time (denoted again t_m) is straightforward. As with all survival distributions, the exponential survival function requires the median value to be the value that makes the survival function equal to one-half. Specifically,

$$S(t_m) = \tfrac{1}{2} = e^{-\lambda t_m} \quad \text{or} \quad t_m = \log(2)/\lambda = \mu \log(2).$$

The natural estimate of the median value $\hat{t}_m$ is then $\log(2)/\hat{\lambda}$ or $\hat{\mu}\log(2)$ and is the maximum likelihood estimate because it is a function of the maximum likelihood estimate $\hat{\mu}$. Therefore, the estimated median survival time has the properties of a maximum likelihood estimate, namely an approximate normal distribution with minimum variance when the sample size is large. From the SFMHS data, the estimated median survival time is $\hat{t}_m = \log(2)/0.043 = 23.084 \log(2) = 16.005$ months.

The accuracy of the estimated confidence interval for the median survival time is also improved by a logarithmic transformation. Specifically, the transformation $\log(\hat{t}_m)$ is used and the estimated variance of the distribution of the logarithm of the estimated median $\hat{t}_m$ is

$$\begin{aligned} \text{variance}[\log(\hat{t}_m)] &= \text{variance}(\log[\hat{\mu}\log(2)]) \\ &= \text{variance}(\log(\hat{\mu}) + \log[\log(2)]) \\ &= \text{variance}[\log(\hat{\mu})] = \frac{1}{d}. \end{aligned}$$

The variance of the logarithm of the median value is the same as the variance of the logarithm of the mean value because adding a constant to a variable does not change its variance (details are given at the end of the chapter). Therefore, the width of a confidence-interval-based on the estimated median value is smaller than a confidence interval based on the estimated mean value [the ratio of the lengths is $\log(2)$]. For normally distributed data, the estimated median value is always less precise than the estimated mean value (larger confidence interval).

Parallel to the estimated mean value, an approximate 95% confidence interval for the logarithm of the median survival time of the population that produced the SFMHS data is then $\log(16.001) \pm 1.960\sqrt{1/155} = 2.773 \pm 0.157$, yielding the confidence interval $(2.615, 2.930)$. The approximate 95% confidence interval for the median survival time estimated by $\hat{t}_m = 16.001$ becomes $(e^{2.615}, e^{2.930}) = (13.670, 18.729)$.

More succinctly, because the median value is $\log(2)\hat{\mu} = 16.001$, the lower and upper bounds of a 95% confidence interval are the same function of the confidence interval bounds based on the estimated mean survival

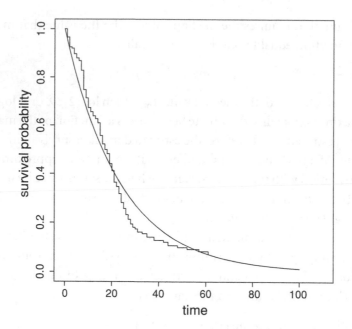

Figure 5.4.   Estimated product-limit survival probabilities (step function) and exponential survival
function for the SFMHS data (continuous line).

time. Specifically, the approximate 95% confidence interval bounds are again
$\log(2)[19.721] = 13.670$ and $\log(2)[27.020] = 18.729$. The values 19.721
and 27.020 are the previous 95% confidence bounds based on the estimated
mean survival time $\hat{\mu}$. As mentioned, the length of the confidence interval
for the median survival time is smaller than the interval for the mean by
a factor of $\log(2) = 0.693$, where $18.729 - 13.607 = 5.059$ versus $27.020 -
19.721 = 7.298$, respectively.

A critical question that must be addressed as part of an analysis based
on a parametric survival distribution is: Does the postulated exponential
probability distribution accurately represent the relationship between time
and the likelihood of survival? or, Is the exponential model useful? or, Is the
hazard rate constant? An investigation of this issue (goodness-of-fit) begins
with the comparison of the product-limit estimated survival probabilities
(no parametric assumption) to the corresponding estimated exponential
survival probabilities (parametric assumption). For the SFMHS data, Figure 5.4 displays this comparison where the natural parametric estimate of
the exponential survival function is

$$\hat{S}(t) = e^{-\hat{\lambda}t} = e^{-0.043t}.$$

The graphic comparison of nonlinear relationships is not as simple or intuitive as the comparison of straight lines. For the exponential survival function, a transformation produces a straight line. Specifically, the transformation $\log(-\log[S(t)])$ yields the straight line, denoted $l(t)$,

$$l(t) = \log(-\log[S(t)]) = \log(-\log[e^{-\lambda t}]) = \log(\lambda t) = \log(\lambda) + \log(t).$$

Thus, the "log–log" transformed estimated survival probabilities from an exponential survival time distribution randomly deviate from a straight line with intercept $\log(\lambda)$ and slope 1.0. Figure 5.5 illustrates the log–log transformation applied to two sets of fictional survival time data. The plot labeled "exponential" is a random sample from an exponential distribution and the other is not. Estimated survival probabilities transformed into straight lines clearly indicate the correspondence (top figure) and the lack of correspondence (bottom figure) between theory and data.

For the San Francisco AIDS patients ($n = 174$), the estimated straight line based on an exponential distribution is $\hat{l}(t_i) = \log(-\log[\hat{S}(t_i)]) = -3.139 + \log(t_i)$ with intercept $= \log(0.043) = -3.319$ and slope $= 1.0$ (solid line in Figure 5.6). Applying the same transformation to the product-limit estimated survival probabilities $\hat{P}_i$ produces values $\log(-\log[\hat{P}_i])$. The least-squares-estimated straight line based on these transformed survival probabilities and the values $\log(t_i)$ yields a straight line with estimated intercept -3.696 and slope 1.195 (dashed line in Figure 5.6). The estimated line $-3.696 + 1.195 \log(t_i)$ provides a second linear measure of goodness-of-fit that does not depend on postulating an underlying survival distribution (model-free). That is, the slope and intercept can have any value. The correspondence between the exponential model and the distribution-free product-limit estimated survival probabilities is now clearly displayed as a difference between two straight lines (Figure 5.6). Both lines randomly differ from the same "45° line" by chance alone when the survival data are sampled from an exponential probability distribution (constant hazard rate). Substantial differences provide clear visual evidence that the survival times are not adequately described by a single parameter exponential distribution, which does not appear to be the case for the SFMHS data.

A survival probability for a selected time t_0 is estimated by $\hat{S}(t_0) = e^{-\hat{\lambda} t_0}$. Similar to the previous confidence intervals, confidence bounds for a specific survival probability are constructed from a transformed estimated value, namely the straight line $l(t)$. The estimate of $l(t)$ for a selected survival time

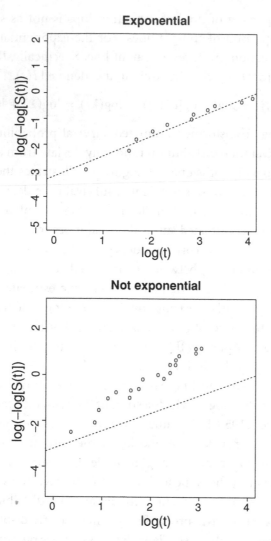

Figure 5.5.   Examples of the fit of transformed exponential and nonexponentially distributed survival times.

t (denoted t_0) has a particularly simple variance and confidence interval. When $\hat{l}(t) = \log(-\log[\hat{S}(t)]) = \log(\hat{\lambda}) + \log(t_0)$, the estimated variance of the distribution of the estimated value $\hat{l}(t_0)$ is

$$\text{variance}(\hat{l}[t_0]) = \text{variance}[\log(-\log[\hat{S}(t_0)])] = \text{variance}[\log(\hat{\lambda}) + \log(t_0)]$$
$$= \text{variance}[\log(\hat{\lambda})] = \frac{1}{d}.$$

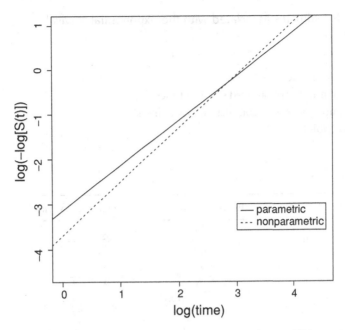

Figure 5.6. Comparison of transformed survival curves (goodness-of-fit).

Because t_0 is a selected value, it is, therefore, a nonrandom (fixed) value. Its variance is zero (variance$[\log(t_0)] = 0$). For example, suppose interest is focused on the survival time $t_0 = 20$ months, then the estimated survival probability $\hat{S}(t_0) = \hat{S}(20) = e^{-\hat{\lambda}(20)} = e^{-0.043(20)}$ is 0.420 and the transformed value is $\hat{l}(t_0) = \hat{l}(20) = \log(0.043) + \log(20) = -0.143$. The normal distribution derived approximate lower and upper 95% confidence interval bounds based on $\hat{l}(20)$ are

$$\text{lower bound} = \hat{l}(t_0) - 1.960\sqrt{1/d} = -0.143 - 1.960\sqrt{1/155} = -0.301$$

and

$$\text{upper bound} = \hat{l}(t_0) + 1.960\sqrt{1/d} = -0.143 + 1.960\sqrt{1/155} = 0.014.$$

The approximately 95% confidence interval bounds based on the survival probability estimated by $\hat{S}(20)$ become

$$\text{lower bound} = e^{-\exp(0.014)} = 0.363$$

and

$$\text{upper bound} = e^{-\exp(0.301)} = 0.477$$

and necessarily $\hat{S}(20)e^{-\exp(0.143)} = 0.420$.

Table 5.1. Estimates associated with the exponential survival distribution.

Notation:

t = survival time,
n = total number of sampled survival times,
d = number of deaths (complete survival times),
p = probability,

and

$i = 1, 2, \ldots, n.$

Parameter	Symbol	Estimated value	Estimated variance
Mean	μ	$\hat{\mu} = \dfrac{\sum t_i}{d}$	$S_{\hat{\mu}}^2 = \dfrac{\hat{\mu}^2}{d}$
Rate	λ	$\hat{\lambda} = \dfrac{d}{\sum t_i}$	$S_{\hat{\lambda}}^2 = \dfrac{\hat{\lambda}^2}{d}$
Median	t_m	$\hat{t}_m = \dfrac{-\log(0.5)}{\hat{\lambda}}$	$S_{\hat{t}_m}^2 = \dfrac{\hat{t}_m^2}{d}$
Percentile	t_p	$\hat{t}_p = \dfrac{-\log(1-p)}{\hat{\lambda}}$	$S_{\hat{t}_p}^2 = \dfrac{\hat{t}_p^2}{d}$

Note:

$\text{variance}(\log[\hat{\mu}]) = \text{variance}(\log[\hat{\lambda}]) = \text{variance}(\log[\hat{t}_{0.5}])$

$= \text{variance}(\log[\hat{t}_p]) = \text{variance}(\log(-\log[\hat{S}(t_0)])) = \dfrac{1}{d}.$

because the five estimates differ only by constant values.

One last point: the linear relationship $l(t) = \log(-\log[S(t)]) = \log(\lambda) + \log(t)$ suggests a more general survival model created by postulating a slope other than 1.0. In symbols, this model is

$$l(t) = \log(-\log[S(t)]) = \log(\lambda) + \gamma \log(t) \qquad (\text{slope} = \gamma).$$

Solving this expression for $S(t)$ yields the survival function $S(t) = e^{-\lambda t^\gamma}$. This two-parameter ($\lambda$ and γ) survival function, called the *Weibull survival function*, is the next topic (Chapter 6). The data in Figure 5.5 (bottom) are an example of log–log transformed survival times sampled from a Weibull distribution (slope = γ = 1.8). Table 5.1 contains a review of the estimates and variances from the exponential survival distribution.

Summarizing mortality and disease rates

The best way to describe the differences between two or more sets of mortality rates is a direct comparison of age-specific rate to age-specific rate. When the number of age-specific rates is large or several populations are compared, a direct "rate-to-rate" comparison remains important but lacks the simplicity of comparing summary values. For example, overall mortality risk is frequently described by combining a sequence of age-specific mortality rates into a single summary age-adjusted mortality "rate." Direct and indirect age-adjustment are popular techniques that produce "adjusted rates" from age-specific mortality or disease data ([1] or [4]).

The easily calculated and interpreted probability of death also serves as a summary of age-specific rates. To illustrate, Table 5.2 contains the number of deaths (d_x) from lung cancer for males and females in the United States (year 2000) and the corresponding number of individuals at risk (l_x) for age x. Also included are the sex- and age-specific estimated average mortality rates per 100,000 person-years calculated in the usual way (denoted $\hat{\lambda}_x$). Implicit is the assumption that these male and female estimated mortality rates are at least approximately constant within each of the relatively short ten-year age intervals. This assumption allows the direct calculation of the age-specific survival probabilities based on the exponential survival distribution (constant hazard rate). These probabilities can then be combined to estimate the survival probability.

To summarize the age-specific rates λ_x with a single survival probability, each rate is converted into an age-specific conditional survival probability p_x based on the assumption that the rate is constant within each age interval x to $x + \delta$ (exponential survival time). The age-specific exponential survival probabilities become

$$p_x = P(\text{surviving from } x \text{ to } x + \delta) = e^{-\hat{\lambda}_x \delta}.$$

For example, the male lung cancer rate 62.7/100,000 for ages 50 to 60 translates into the conditional survival probability $P(\text{surviving from 50 to 60}) = \hat{p}_{60} = e^{-0.000627(10)} = 0.9938$ ($\delta = 10$ years).

Much like the product-limit and life table estimates, the product of the age-specific conditional survival probabilities succinctly summarize the mortality risk over a sequence of age intervals. From Table 5.2, the probability of dying

Table 5.2. Age-specific rates of lung cancer deaths for U.S. males and females.

	Males			Females		
Age	Deaths (d_x)	At-risk (l_x)	Rate (λ_x)*	Deaths (d_x)	At-risk (l_x)	Rate (λ_x)*
50–60	9915	15,819,750	62.7	6919	16,616,620	41.7
60–70	21,218	9,821,436	216.0	14,239	11,012,221	129.3
70–80	29,280	7,280,991	402.1	21,069	9,559,703	220.4
80–90	13,754	3,110,375	442.2	11,808	5,529,059	213.6
90+	1394	454,171	392.6	1783	1,346,803	132.4
Total	75,561	36,486,723	207.1	55,818	44,064,406	126.7

* Age-specific lung cancer mortality rates per 100,000 person-years, ages x to $x + 10$ years (age = $x \geq 50$).
Source: National Center for Health Statistics, 2000.

of lung cancer after the age of 50 is estimated by

$$P(\text{dying of lung cancer beyond age 50}) = 1 - [\hat{p}_{60} \times \hat{p}_{70} \times \hat{p}_{80} \times \hat{p}_{90+}] =$$
$$1 - [e^{-\hat{\lambda}_{60}\delta} \times e^{-\hat{\lambda}_{70}\delta} \times e^{-\hat{\lambda}_{80}\delta} \times e^{-\hat{\lambda}_{90}\delta} \times e^{-\hat{\lambda}_{60}\delta}] = 1 - [e^{-\delta \sum \hat{\lambda}_x}] \approx \delta \sum \hat{\lambda}_x.$$

The approximation $e^{-x} \approx 1 - x$ is accurate when x is moderately small $(-0.2 \leq x \leq 0.2$—Figure 5.7).

Approximately, the probability of death from a specific cause over a period of time (a sequence of intervals) is the sum of the age-specific rates multiplied by the lengths of the corresponding age intervals ($\sum \delta_i \hat{\lambda}_i$). For males ($\delta_i = 10$ years), a summary probability of dying of lung cancer after age 50 is approximately

$$P(\text{male death}) \approx 10[62.68/100,000 + 216.04/100,000$$
$$+ \cdots + 306.93/100,000] = 0.143.$$

The comparable summary probability from the female lung cancer mortality data is $P(\text{female death}) \approx 0.074$. The straightforward interpretation of these probabilities is that an estimated 14.3% of the males and 7.4% of the females will die of lung cancer after age 50. Both probabilities are "age-adjusted" in the sense that differences in male/female age distributions have essentially no impact on the comparison.

Although it is not usually an issue with national data (Table 5.2) because the extremely large numbers of persons-at-risk makes the variance

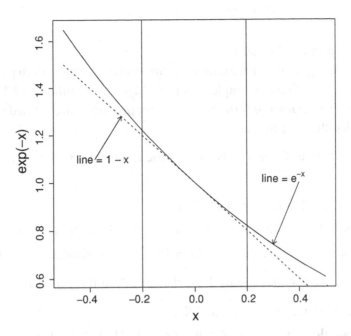

Figure 5.7.   The geometry of approximating e^{-x} by $1 - x$.

inconsequentially small, the estimated variance of the distribution of this estimated survival probability $\hat{P}_k = 1 - \sum \delta_i \, \hat{\lambda}_i$ is approximately

$$\text{variance}(\hat{P}_k) \approx \sum \frac{\delta_i^2 \, d_i}{l_i^2} \qquad i = 1, 2, \ldots, k = \text{number age intervals.}$$

The symbol d_i represents the number of deaths in the ith age interval of length δ_i among l_i individuals at risk. For the U.S. lung cancer mortality data, the standard errors are $\sqrt{\text{variance}(\hat{P}_{\text{male}})} = 0.00095$ and $\sqrt{\text{variance}(\hat{P}_{\text{female}})} = 0.00042$, showing the negligible influence from sampling variation on these estimated summary probabilities due to the large number of observations. Unlike the estimate of the summary survival probability $\hat{P}_k$, the variance of the estimate $\hat{P}_k$ is influenced by differences in the age distributions of the groups compared.

These estimated survival probabilities are approximate for two reasons. First, the approximate value $(1 - \sum \delta_i \lambda_i)$ is used instead of the exact value $(e^{-\sum \delta_i \lambda_i})$. Second, and of more importance, cross-sectional data (the year 2000) are used to estimated probabilities over a 50-year-period, which would be strictly correct only if mortality risk remained perfectly constant.

APPENDIX

Statistical tools: properties of variance

The following reviews three relevant properties of variance (both population variance and estimated sample variance). Rigorous proofs are readily found in elementary statistical texts. Here the three properties are defined and small examples illustrate them.

1. The variance of $X + a$ is the same as the variance of X when a is a constant or

   variance$(X + a) =$ variance(X) ·

   Example. X: {2, 4, 6, 8, 10}   and $X + 5$: {7, 9, 11, 13, 15}

   where $a = 5$: variance$(X + 5) = 10$ and   variance$(X) = 10$.

2. The variance of aX equals the variance of X multiplied by a^2 or

   variance$(aX) = a^2$ variance (X).

   Example. X: {2, 4, 6, 8, 10} and aX: {1, 2, 3, 4, 5} where $a = \frac{1}{2}$:

   variance$(aX) = 2.5$ and a^2 variance$(X) = \left(\frac{1}{2}\right)^2 10 = \dfrac{10}{4} = 2.5$.

3. The variance of $X + Y$ equals variance(X) plus variance(Y) when X and Y are uncorrelated, or

   variance$(X + Y) =$ variance$(X) +$ variance(Y)

   when correlation $(X, Y) = 0$.

   Example. X: {2, 4, 6, 8, 10}   and Y: {1, 7, 14, 7, 1}:

   variance$(X) = 10$ and variance$(Y) = 29$, making variance$(X + Y) = 39$,

   because *correlation* $(X, Y) = 0$.

   In general, for values $x_1, x_2, \ldots, x_k$, the variance of a sum is the sum of the variances,

$$\text{variance}\left(\sum x_i\right) = \sum \text{variance}(x_i),$$

when correlation $(x_i, x_j) = 0$ for all $i \neq j = 1, 2, \ldots, k$.

Weibull survival time probability distribution

Postulating that survival times are sampled from a population with a constant hazard rate is undoubtedly unrealistic in many situations. The Weibull probability distribution is defined by two parameters; a parameter (denoted $\lambda, \lambda \geq 0$) called the *scale parameter* and a parameter (denoted $\gamma, \gamma \geq 0$) called the *shape parameter*. The two parameters of the Weibull probability distribution provide increased flexibility that potentially improves the description of collected survival time data. Of most importance, the shape parameter allows the hazard function to increase or decrease with increasing survival time.

To repeat (Chapter 5), the Weibull survival probabilities are given by the expression

$$S(t) = P(T \geq t) = e^{-\lambda t^{\gamma}}.$$

Using two algebraic approximations, the Weibull distribution hazard function can be justified from the previous definition of a relative rate (Chapter 1). The approximations are

$$(t + \delta)^g = t^g + gt^{g-1}\delta + \tfrac{1}{2}g(g - 1)t^{g-2}\delta^2 + \cdots + \delta^g \approx t^g + gt^{g-1}\delta$$

for small values of $\delta(\delta < 0.2)$, and, again

$$e^{-x} \approx 1 - x \quad \text{(Chapter 5)}.$$

For example, when $t = 20$, $\delta = 0.1$, and $g = 1.8$, $(t + \delta)^g = (20 + 0.1)^{1.8} = 221.693$ and approximately $t^g + gt^{g-1}\delta = 20^{1.8} + 1.8(20^{0.8})(0.1) = 221.690 \approx 221.693$. Then, for small values of δ, the hazard function for the

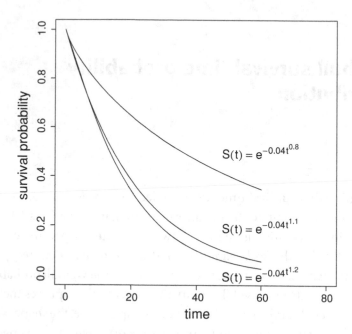

Figure 6.1.   Three Weibull survival functions $S(t)$(scale $= \lambda = 0.04$, shape $= \gamma = \{0.8, 1.1, \text{and } 1.2\}$).

Weibull distribution is

$$h(t) = -\frac{\frac{d}{dt}S(t)}{S(t)} \approx -\frac{[S(t+\delta) - S(t)]/\delta}{S(t)}$$

$$= -\frac{[e^{-\lambda(t+\delta)^\gamma} - e^{-\lambda t^\gamma}]/\delta}{e^{-\lambda t^\gamma}} = -\frac{e^{-\lambda(t+\delta)^\gamma + \lambda t^\gamma} - 1}{\delta}$$

$$\approx -\frac{1 - \lambda[(t+\delta)^\gamma - t^\gamma] - 1}{\delta} = \frac{\lambda[t^\gamma + \delta\gamma t^{\gamma-1} - t^\gamma]}{\delta} = \lambda\gamma t^{\gamma-1}.$$

The identical hazard function also follows directly from the general definition of an instantaneous relative rate (Chapter 1), or the hazard rate is

$$h(t) = -\frac{\frac{d}{dt}S(t)}{S(t)} = \frac{\lambda\gamma t^{\gamma-1}e^{-\lambda t^\gamma}}{e^{-\lambda t^\gamma}} = \lambda\gamma t^{\gamma-1}.$$

Figures 6.1 and 6.2 display the two-parameter Weibull survival and hazard functions ($\lambda = 0.04$ and $\gamma = 0.8, 1.1,$ and 1.2). For $\gamma = 1.0$, the Weibull and the exponential survival probabilities are identical (constant hazard).

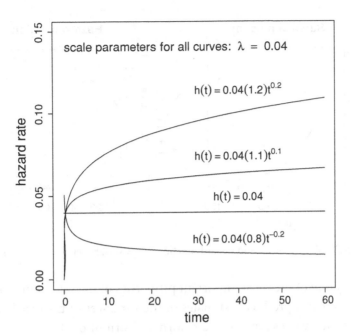

Figure 6.2.   Four Weibull hazard functions $h(t)$(scale $= \lambda = 0.04$, shape $= \gamma = \{0.8, \ 1.0, \ 1.1,$ and $1.2\}$).

For $\gamma > 1.0$, the hazard rate strictly increases in a nonlinear pattern with increasing survival time. Human mortality and disease patterns, for example, have increasing hazard rates with age. For $\gamma > 1.0$, the hazard rate strictly decreases, also in a nonlinear pattern, with increasing survival time. For example, the risk of a recurrence of a tumor after surgery might be characterized by a decreasing hazard rate with increasing time. Further illustrations of Weibull survival and hazard functions are displayed at the end of the chapter (see Figures 6.10, 6.11, and 6.12).

To repeat from Chapter 3, the maximum likelihood estimates of the two defining parameters of the Weibull survival distribution are the most efficient estimates (lowest variances) and have approximate normal distributions. However, the estimation of these two parameters is numerically complex and is usually done with a computer program. The maximum likelihood estimation process also accounts for noninformative censored observations making the estimates unbiased. As mentioned, the bias incurred from censored values is rarely a concern once it is accounted for in the estimation process.

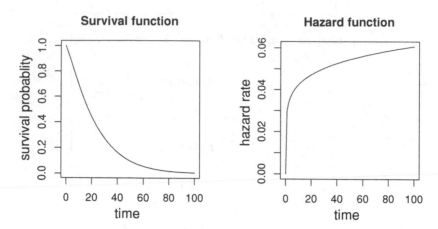

Figure 6.3.   Estimated Weibull survival and hazard functions ($\lambda = 0.0257$ and $\gamma = 1.156$).

Certainly one of the most important properties of the maximum likelihood process is that it produces estimates of the variances of the distributions of the estimated values (approximate normal distributions).

For the SFMHS AIDS survival data ($n = 174$), the computer-generated maximum likelihood estimates of the Weibull distribution defining parameters λ and γ and their associated variances are as follows:

scale parameter: $\hat{\lambda} = 0.0257$ with estimated variance$(\hat{\lambda}) = 0.00004$

and

shape parameter: $\hat{\gamma} = 1.156$ with estimated variance$(\hat{\gamma}) = 0.00519$.

Estimates of the survival and hazard functions follow directly from the estimated parameters (Figure 6.3). They are as follows:

estimated survival function: $\hat{S}(t) = e^{-\hat{\lambda}t^{\hat{\gamma}}} = e^{-0.0257t^{1.156}}$

and

estimated hazard function: $\hat{h}(t) = \hat{\lambda}\hat{\gamma}\,t^{\hat{\gamma}-1} = 0.0257(1.156)t^{0.156}$.

Confidence intervals based on each of the estimated Weibull survival function parameters are again more accurately constructed from a logarithmic transformation. The estimates $\hat{\lambda}$ and $\hat{\gamma}$ from 174 AIDS patients illustrate.

A log transformation yields

$$\log(\hat{\lambda}) = -3.662 \text{ with a estimated variance}[\log(\hat{\lambda})]$$
$$= \frac{1}{\hat{\lambda}^2} \text{variance}(\hat{\lambda}) = 0.0663$$

and

$$\log(\hat{\gamma}) = 0.145 \text{ with an estimated variance}[\log(\hat{\gamma})]$$
$$= \frac{1}{\hat{\gamma}^2} \text{variance}(\hat{\gamma}) = 0.00388.$$

The variances of $\hat{\lambda}$ and $\hat{\gamma}$, as noted, are estimated as part of the maximum likelihood estimation process. An approximate 95% confidence interval based on the normal distribution and the transformed log-estimate is $\log(\text{estimate}) \pm 1.960\sqrt{\text{variance}[\log(\text{estimate})]}$. As before, the log transformation improves the normal distribution approximation by creating a more symmetric distribution and exponentiating guarantees that the lower limit is positive. Specifically, for $\log(\lambda)$, the 95% approximate confidence interval is

$$\text{bounds} = -3.662 \pm 1.960\sqrt{0.0663} \quad \text{or} \quad (A, B) = (-4.167, -3.157),$$

and for $\log(\gamma)$, it is

$$\text{bounds} = 0.145 \pm 1.960\sqrt{0.0039} \quad \text{or} \quad (A, B) = (0.023, 0.267).$$

Exponentiating these estimated bounds (e^A, e^B) yields approximate 95% confidence intervals for the Weibull parameters. They are as follows:

scale parameter: $\hat{\lambda} = 0.0257$ produces the 95% confidence interval

$$(0.016, 0.043)$$

and

shape parameter: $\hat{\gamma} = 1.156$ produces the 95% confidence interval

$$(1.023, 1.306).$$

The log–log transformation introduced in Chapter 5 provides alternative and intuitive estimates of the two Weibull distribution parameters because the transformed Weibull survival probabilities are random deviations from a straight line. Again, the line is

$$l_i = \log(-\log[\hat{P}_i]) = a + b\log(t_i) = \log(\lambda) + \gamma\log(t_i),$$

where the product-limit survival probabilities $\hat{P}_i$ are estimated from the observed survival times t_i. Usual least-squares estimation yields estimates of the intercept $\hat{a}$ and the slope $\hat{b}$ and, therefore, the Weibull parameters estimates of $\hat{\lambda} = e^{\hat{a}}$ and $\hat{b} = \hat{\gamma}$.

For the previous SFMHS data ($n = 172$ and $d = 155$), the product-limit estimated survival probabilities $\hat{P}_i$ produce the line $\hat{l}_i = -3.696 + 1.195\log(t_i)$ from the 155 observed pairs $= \{x_i, y_i\} = \{\log(t_i), \log(-\log[\hat{P}_i])\}$. The Weibull parameter estimates $\hat{\lambda} = e^{-3.696} = 0.025$ (scale) and $\hat{\gamma} = 1.195$ (shape) follow directly. The previous maximum likelihood estimates ($\hat{\lambda} = 0.0257$ and $\hat{\gamma} = 1.156$) have optimum properties and are almost always used. However, estimates based on a straight line are conceptually simple and provide a natural alternative to the more complex computer-generated maximum likelihood estimates.

The estimated mean survival time based on the conjecture of a Weibull survival distribution is

$$\hat{\mu} = \hat{\lambda}^{-1/\hat{\gamma}}\Gamma(1 + 1/\hat{\gamma}).$$

The symbol $\Gamma(1 + 1/\gamma)$ represents a gamma function evaluated at $1 + 1/\gamma$. (Γ is the capital Greek letter gamma.) A plot of the gamma values $\Gamma(x)$ for $x = 0.3$ to $x = 3.5$ (the usual range of γ for human survival data) is displayed in Figure 6.4. The gamma function and the derivation of the mean survival time expression are mathematically sophisticated and are not discussed in detail. However, the estimation of the mean value is straightforward because values of a gamma function can be looked up in tables, found on the Internet, or calculated with computer routines. It is worth noting that when $\gamma = 1.0$, the estimated mean survival time is $\hat{\mu} = 1/\hat{\lambda}$ (constant hazard) because $\Gamma(2) = 1.0$. For the AIDS data, the estimated parameters $\hat{\lambda} = 0.0257$ and $\hat{\gamma} = 1.156$, along with the corresponding gamma value $\Gamma(1 + 1/1.156) = \Gamma(1.865) = 0.950$, produce an estimated mean survival time of $\hat{\mu} = 0.0257^{-1/1.156}(0.950) = 22.597$ months.

Estimates of the median survival time or other percentiles of a Weibull probability distribution follow from the estimated survival function $\hat{S}(t)$.

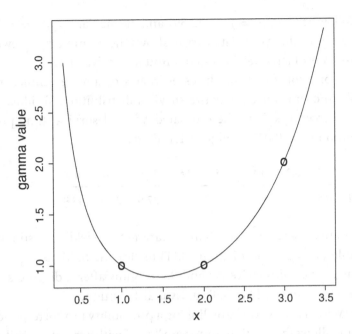

Figure 6.4. Gamma values displayed for x from 0.3 to 3.5 (the usual range for survival data) with
$\Gamma(x = 1) = 1, \Gamma(x = 2) = 1,$ and $\Gamma(x = 3) = 2$ (symbol $= 0$).

For the probability p, the corresponding p-level percentile (denoted t_p) is
estimated by

$$\hat{t}_p = \left[\frac{-\log(1 - p)}{\hat{\lambda}}\right]^{1/\hat{\gamma}}.$$

The technical/mathematical name for this expression is the *inverse Weibull
survival function*. That is, the Weibull survival function $S(t)$ produces a
probability p from a survival time t. The inverse Weibull function produces
a survival time t from a probability p. The percentile expression is found
by solving the relationship $S(\hat{t}_p) = 1 - p$ for the value $\hat{t}_p$. Applied to the
AIDS survival data, the estimated median survival time ($p = 0.5$—the 50th
percentile) is

$$\hat{t}_{0.5} = \left[\frac{\log(2.0)}{0.0257}\right]^{1/1.156} = 17.315 \text{ months.}$$

A rough measure of the symmetry of a distribution of data is the difference
between the mean and the median values. The Weibull summary estimates

are as follows: the mean $= \hat{\mu} = 22.597$ and the median $= \hat{t}_{0.5} = 17.315$. The distribution of the AIDS data clearly shows the asymmetry (skewed to the right—excess of large values) often found in survival data.

The expression for $\hat{t}_p$ produces estimates of any percentile, making it possible to characterize the entire survival distribution. To illustrate, nine selected percentiles from the estimated Weibull survival time probability distribution ($\hat{\lambda} = 0.0257$ and $\hat{\gamma} = 1.156$) are

p	0.1	0.2	0.3	0.4	0.5	0.6	0.7	0.8	0.9
$\hat{t}_p$	3.4	6.5	9.7	13.3	17.3	22.0	27.9	35.9	48.9

For example, the estimated Weibull parameters yield the estimated 30th percentile ($p = 0.3$) of 9.7 months. Thus, for the SFMHS population, an estimated 70% will survive beyond 9.7 months after a diagnosis of AIDS, based on the estimated Weibull survival distribution.

The Weibull inverse function linking a probability to a corresponding survival time allows the creation of random "data" with a specified Weibull probability distribution. For parameters (or estimates of parameters) λ and γ, a randomly generated probability p substituted into the expression for a percentile (inverse function) produces a randomly generated Weibull distribution "survival time." For example, for the estimates $\hat{\lambda} = 0.256$ and $\hat{\gamma} = 1.156$, the randomly generated probability 0.320 produces the randomly generated random survival time $t = [-\log(1 - 0.320)/0.0257]^{1/1.156} = 10.426$. Most computer systems, and even some handheld calculators, produce random probabilities (uniformly distributed values between 0 and 1). Such computer-generated Weibull-distributed "data" are then available to verify theories or estimate variances of complicated functions of the parameters, explore empirically the properties of summary values, or describe just about any property of the Weibull or exponential ($\gamma = 1$) survival distributions. For example, generating a large number of random "survival times" directly produces an accurate estimate of the Weibull-distributed survival times, its mean survival time μ, its median survival time t_m, and their variances without theory or mathematics. The mean value of such a computer-generated distribution based on the AIDS survival data (Weibull parameters $\hat{\lambda} = 0.0257$ and $\hat{\gamma} = 1.156$) is displayed in Figure 6.5. The estimated mean

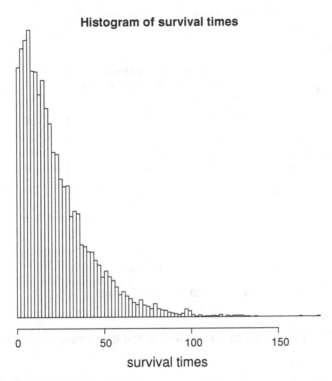

Figure 6.5.   Computer-simulated distribution of 10,000 randomly sampled survival times from a Weibull probability distribution ($\hat{\lambda} = 0.0257$ and $\hat{\gamma} = 1.156$).

value is $\hat{\mu} = \bar{t} = 22.643$ with an estimated variance of the distribution of mean survival times of variance($\hat{\mu}$) $= 2.490$ using 10,000 random "survival times." An estimate of the variance, as always, leads to statistical tests and confidence interval bounds. Similarly, from the same computer-generated data, the estimated median survival time is $\hat{t}_m = 17.048$ with an estimated variance of variance($\hat{t}_m$) $= 3.024$.

Parallel to applying an exponential probability distribution to summarize survival time data, the adequacy of the Weibull survival time distribution as a description of the sampled data should be explored as part of a complete analysis. A goodness-of-fit assessment follows the same pattern suggested for the exponential distribution (Chapter 5).

The comparison of the parametrically estimated Weibull and the non-parametrically (product-limit) estimated survival probabilities is again the

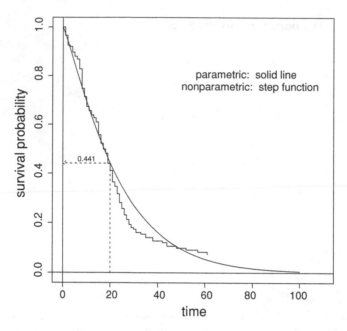

Figure 6.6.   Goodness-of-fit: two survival curves (parametric and model-free) for the SFMHS AIDS data.

first step. The estimated model-generated Weibull survival probabilities (parameters $\hat{\lambda} = 0.0257$ and $\hat{\gamma} = 1.156$—solid line) and the corresponding model-free estimated product-limit survival probabilities (step function) from the SFMHS cohort ($n = 174$) are displayed in Figure 6.6. From the example, for survival time $t = 20$, the Weibull estimated survival probability $P(T \geq 20) = \hat{S}(20) = e^{-0.0257(20)^{1.156}} = 0.442$ and the corresponding product-limit estimate of 0.442 are almost identical.

Again, parallel to the evaluation of the exponential distribution, transformed values of both the Weibull and product-limit estimated survival probabilities improve the visual goodness-of-fit comparison by contrasting straight lines. The Weibull distribution estimated line is $\hat{l}(t) = \log(-\log[\hat{S}(t)]) = \log(\hat{\lambda}) + \hat{\gamma} \log(t) = -3.662 + 1.156 \log(t)$. The product-limit estimated straight line (the ordinary least-squares estimate) is again based on the pairs of values $\log(t_i)$ and $\log(-\log[\hat{P}_i])$, where $\hat{P}_i$ represents a product-limit estimated survival probability. The model-free estimated line from the SFMHS data is $-3.696 + 1.195 \log(t_i)$. Both straight lines are displayed in Figure 6.7. The comparison leaves little doubt

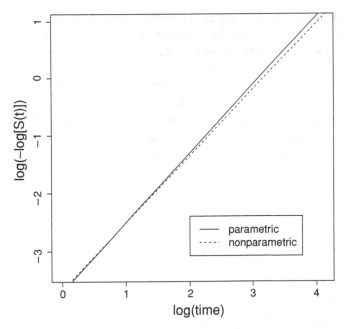

Figure 6.7.　Goodness-of-fit: comparison of the log-log transformed survival functions for the Weibull model (parametric) and the product-limit estimates (model-free).

that the two-parameter Weibull distribution based on parameter estimates $\hat{\lambda} = 0.0257$ and $\hat{\gamma} = 1.156$ closely corresponds to the less parsimonious but model-free product-limit estimate. Thus, the estimated two-parameter Weibull model allows an extremely efficient, useful, and, most important of all, simple description of the survival pattern that the data were collected to describe. Parameters for the three straight lines (intercepts and slopes) calculated from the transformed exponential, product-limit, and Weibull estimated survival probabilities are summarized in Table 6.1 (SFMHS data).

Instead of comparing the parametric exponential or Weibull survival probabilities to the model-free product-limit estimates, an alternative is based on parametric and nonparametric estimated cumulative hazard functions. The cumulative hazard function estimated from the product-limit estimated survival probabilities describes the survival pattern of the $n = 174$ AIDS patients (Figure 6.8; step function—solid line). In symbols, this model-free estimate is $\hat{H}(t_i) = -\log(\hat{P}_i)$ for each complete survival time $i = 1, 2, \ldots, d$. Because the estimated cumulative hazard function is derived from the model-free

Table 6.1. Intercepts and slopes of summary straight lines representing the transformed exponential, product-limit, and Weibull survival distributions (SFMHS data).

	Intercepts	Slopes
Exponential distribution*	$\log(\lambda)$	1.0
Exponential model estimates*	−3.139	1.0
Product-limit estimates	−3.696	1.195
Weibull distribution	$\log(\lambda)$	γ
Weibull model estimates	−3.662	1.156

* From Chapter 5.

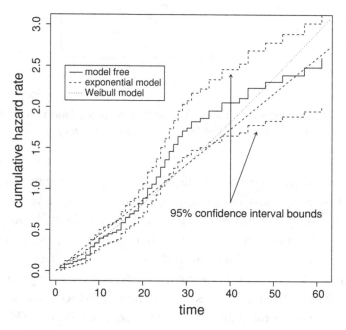

Figure 6.8.   Cumulative hazard functions: model-free product-limit, exponential distribution ($\hat{\lambda} =$ 0.043 and $\gamma = 1.0$), and Weibull distribution ($\hat{\lambda} = 0.0257$ and $\hat{\gamma} = 1.156$) estimates.

product-limit estimate of survival probabilities, it too is model-free. The approximate 95% pointwise confidence interval bounds for product-limit generated survival probabilities (step-function) are

$$\hat{H}(t_k) \pm 1.960\sqrt{\text{variance}[\hat{H}(t_k)]}$$

using the estimated variance of the estimated cumulative hazard function at the time t_k, given by

$$\text{variance}[\hat{H}(t_k)] = \sum \frac{\hat{q}_i}{n_i \hat{p}_i} \qquad i = 1, 2, \ldots, k.$$

This expression for the estimated variance of $\hat{H}(t_k)$ is a variation of the Greenwood variance of an estimated survival probability (Chapters 3 and 4).

In addition, the theoretical (model-based) Weibull cumulative hazard function is

$$H(t) = -\log(S[t]) = -\log(e^{-\lambda t^\gamma}) = \lambda t^\gamma.$$

From the AIDS data ($\hat{\lambda} = 0.0257$ and $\gamma = 1.156$), the estimate becomes $\hat{H}(t) = 0.0257 t^{1.156}$. It follows that the cumulative hazard function for the exponential distribution is $H(t) = \lambda t$ ($\gamma = 1$), a straight line with intercept equal to zero and slope equal to the constant hazard rate λ. From the AIDS data ($\lambda = 0.043$—Chapter 5), the estimate becomes $\hat{H}(t) = 0.043t$. Figure 6.8 displays these two parametrically estimated cumulative hazard rates (Weibull = dotted line and exponential = dashed line). Again, it is visually apparent that the Weibull distribution is an accurate representation of the data and is an improvement over the exponential distribution (constant hazard).

The question arises: Is the Weibull survival distribution a better description of the sampled survival data than the exponential survival distribution? The answer is Yes. Improvement in the correspondence between estimates and data (goodness-of-fit) always occurs when a two-parameter model is chosen over a nested one-parameter model. More extensive models always fits better. The important question becomes: Is the improvement greater than would be expected by chance? Comparing a plot of a survival function or cumulative hazard function estimated from a model to the corresponding plot made directly from the data (Figure 6.7) is a start, but when obvious differences do not occur, this question remains unanswered.

A comparison of log-likelihood values (likelihood ratio test), however, provides a statistically rigorous and quantitative comparison of the Weibull and the exponential survival distributions. The two survival functions are

Table 6.2. Estimated summary values from the exponential and Weibull survival time distributions (SFMHS data, $n = 174$).

Parameter	Symbol	Exponential	Weibull
Estimate	$\hat{\lambda}$	0.043	0.026
Standard error	$\sqrt{\text{variance}(\hat{\lambda})}$	0.003	0.007
Estimate	$\hat{\gamma}$	—	1.156
Standard error	$\sqrt{\text{variance}(\hat{\gamma})}$	—	0.072
Mean	$\hat{\mu}$	23.262	22.643*
Standard error	$\sqrt{\text{variance}(\hat{\mu})}$	1.864	1.578*
Median	$\hat{t}_{0.5}$	16.197	17.048*
Standard error	$\sqrt{\text{variance}(\hat{t}_{0.5})}$	1.653*	1.739*

* Estimates from 10,000 simulated values (Figure 6.5).

nested models because the exponential distribution ($\gamma = 1$) is a special case of the Weibull distribution ($\gamma \neq 1$). For the AIDS data, the corresponding log-likelihood values are

$$\log - \text{likelihood} = \log(L_{\gamma=1}) = -641.566$$

$$(\text{exponential distribution} - \gamma = 1)$$

and

$$\log - \text{likelihood} = \log(L_{\gamma \neq 1}) = -639.078$$

$$(\text{Weibull distribution} - \gamma \neq 1),$$

yielding the likelihood ratio test-statistic

$$X^2 = -2[\log(L_{\gamma=1}) - \log(L_{\gamma \neq 1})]$$
$$= -2[-641.566 - (-639.078)] = 4.976.$$

The test statistic X^2 has a chi-square distribution with one degree of freedom when the observed difference reflects only random influences of sampling variation. That is, the difference arises because the two-parameter model always more effectively capitalizes on random variation to produce a better fit and this property is the only source of improvement. The small p-value $P(X^2 \geq 4.976 \mid \gamma = 1) = 0.026$ indicates that random variation is not likely the entire explanation for the difference in log-likelihood values. It is therefore likely that the additional flexibility of the Weibull model takes advantage of a systematic pattern within the AIDS data to more accurately represent the survival pattern. More simply, evidence exists that the hazard rate is not constant (γ is not equal to one). This result is consistent with the previous 95% confidence interval (1.0024, 1.306) based on the estimated scale parameter $\hat{\gamma} = 1.156$. The confidence interval does not contain the value 1.0. From both perspectives, it appears that an increasing hazard function reflects an important aspect of the SFMHS survival data not captured by an exponential model (constant hazard). The estimated values for both the exponential and Weibull survival time distributions are summarized in Table 6.2.

APPENDIX

Statistical tools: properties of inverse functions

The mathematical term *inverse function* is familiar in statistics. The standard normal distribution allows a probability to be calculated from a percentile and the inverse function allows a percentile to be calculated from a probability. The inverse standard normal distribution is a complicated function, but a table of standard normal values produces probabilities from percentiles and, therefore, inversely produces percentiles from probabilities. For example, the standard normal distribution produces the probability $p = 0.95$ from the 95th percentile $z = 1.645$ and the inverse normal function produces the 90th percentile $z = 1.282$ from the probability $p = 0.90$.

Some functions have simple and easily derived inverses. For example, for the function $F(x) = x^2$, the inverse function is $G(x) = \sqrt{x}$. The defining property of an inverse function is that $F[G(x)] = x$ or $G[F(x)] = x$. Specifically, $(\sqrt{x})^2 = \sqrt{x^2} = x$.

The exponential function provides another example, where

$$F(t) = e^{-\lambda t} \text{ and } G(p) = -\log(p)/\lambda, \text{ where } F[G(p)]$$
$$= p \text{ and } G[F(t)] = t.$$

Geometrically, the function $F(t) = e^{-t}(\lambda = 1)$ describes the path from the value t to the probability p (Figure 6.9a). From the plot, the value $t = 1.5$ produces the probability $F(1.5) = e^{-}1.5 = 0.223$. The inverse function $G(p)$ describes the path from the probability p to the value t (Figure 6.9b). From the plot, the probability $p = 0.4$ produces the value $G(0.4) = -\log(0.4) = 0.916$.

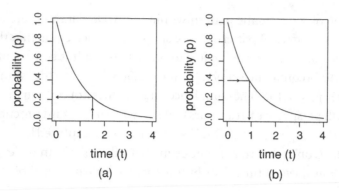

Figure 6.9. (a) $F(t) = \exp(-t)$–function. (b) $G(p) = -\log(p)$–inverse function.

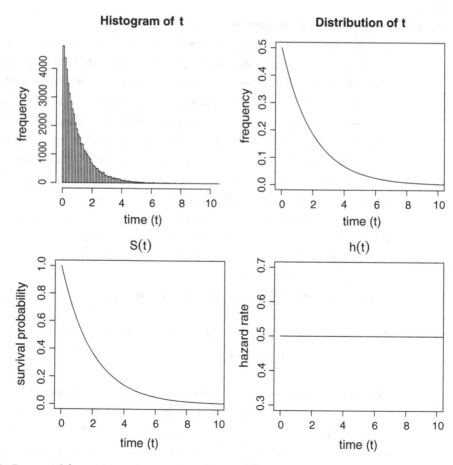

Figure 6.10. Exponential—$\lambda = 0.5$ and $\gamma = 1.0$ or $S(t) = e^{-0.5t}$ and $h(t) = 0.5$.

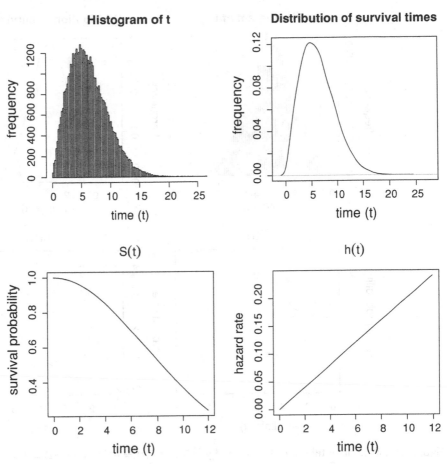

Figure 6.11. Weibull distribution—$\lambda = 0.5$ and $\gamma = 2.0$ or $S(t) = e^{-0.5t^2}$ and $h(t) = 1.0t$.

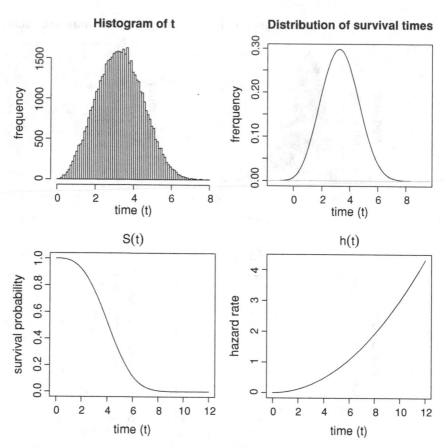

Figure 6.12. Weibull distribution—$\lambda = 0.5$ and $\gamma = 3.0$ or $S(t) = e^{-0.5t^3}$ and $h(t) = 1.5.t^2$.

Analysis of two-sample survival data

Rigorous statistical comparisons between small sets of data (20 observations or so) began with W. S. Gossett's t-distribution (1908). The statistical comparison of data from two groups is fundamentally important in general. Both parametric and nonparametric methods exist to identify differences in survival experience based on data sampled from two populations. The log-rank test illustrates the nonparametric approach and the two-sample hazards model illustrates the parametric approach. In addition, the two-sample hazards model serves as an introduction to the multivariable hazards model (Chapters 8 and 9).

Two-sample analysis: log-rank test

A popular nonparametric two-sample comparison technique, called the *log-rank test*, begins with classifying survival data into a sequence of intervals based on the times of death. Parallel to the product-limit estimation of survival probabilities (Chapter 4), a sequence of 2×2 tables is created from this sequence. For each time interval, the collected observations are classified by binary outcome and risk factor variables. Again avoiding general terminology, the outcome variable is referred to as died/survived and the two compared groups are referred to as risk factor present (F) or absent ($\bar{F}$). However, the log-rank test applies to most kinds of survival or failure time data.

A 2×2 table is created for each unique and complete survival time t_i (each time interval t_{i-1} to t_i). The form of this 2×2 table describing n_i at-risk individuals at survival time t_i is

Time $= t_i$	Dead	Alive	Total
F	a_i	b_i	$a_i + b_i$
$\bar{F}$	c_i	d_i	$c_i + d_i$
Total	$a_i + c_i$	$b_i + d_i$	n_i

Similarly to the analysis of many 2×2 tables, the individuals at risk are classified into these tables to address the question: Is the risk factor associated with survival? To effectively address this question with a single measure of association, the chosen measure must be constant with respect to survival time. To create such a single comprehensive summary, the data are stratified by time of death. A measure of risk is uninfluenced by survival time when it is calculated within each stratum (table) and combined over all strata to summarize the association between risk factor and outcome.

Consider the survival times of $n = 23$ African-American male participants in the SFMHS. These times are months from diagnosis of AIDS to death from AIDS or to the end of study participation (censored). The survival times classified by nonsmoker ($\bar{F}$) and smoker (F) status are as follows:

Nonsmokers ($\bar{F}$): 2^+, 42^+, 27^+, 22, 26^+, 16, 31, 37, 15, 30, 12^+, 5, 80, 29, 13, 1, and 14

Smokers (F): 21^+, 4, 25, 8, 23, and 18.

There are $n_0 = 17$ nonsmokers with five censored survival times and $n_1 = 6$ smokers with one censored survival time.

A good place to start the comparison of the survival experiences between two groups is a plot of the two product-limit estimated survival functions. This plot has the advantage that it contrasts survival times in natural units, allowing a visual and intuitive comparison of group differences. Product-limit estimated survival functions for SFMHS African-Americans nonsmokers and smokers are plotted in Figure 7.1. The disadvantage of a visual comparison is that it does not formally account for the influence of sampling variation on the estimated values. A log-rank test brings this element to a comparison and together both methods produce a distribution-free evaluation of the observed differences in survival experience between two groups of sampled individuals.

The nonsmoking/smoking data produce a sequence of 17 log-rank 2×2 tables. A single table is constructed for each unique and complete survival

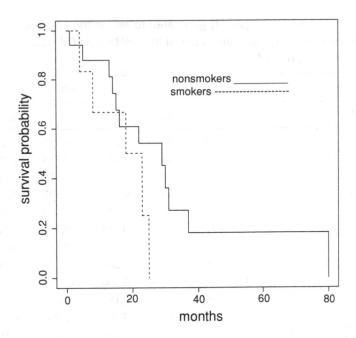

Figure 7.1. Product-limit estimated survival functions for SFMHS African-Americans, comparing non-smokers and smokers.

time. Parallel to product-limit estimation, incomplete survival times are included only in those tables (time intervals) completed by the censored individuals. For the AIDS smoking data, the time of death $t_1 = 1$ month (time interval 0 to 1 month) generates the first 2×2 table, where

$t_1 = 1$	Dead	Alive	Total
Smoker	0	6	6
Nonsmoker	1	16	17
Total	1	22	23

For the second complete survival time, $t_2 = 4$ months (time interval 1 to 4 months), the next 2×2 table is

$t_2 = 4$	Dead	Alive	Total
Smoker	1	5	6
Nonsmoker	0	15	15
Total	1	20	21

Table 7.1. The 17 tables (strata) generated to assess the association between smoking exposure and survival in SFMHS data (African-American males, $n = 23$).

Interval	Data					Estimates		
	a_i	b_i	c_i	d_i	n_i	A_i	C_i	$\hat{v}_i$
0–1	0	6	1	16	23	0.261	0.739	0.193
1–4	1	5	0	15	21	0.286	0.714	0.204
4–5	0	5	1	14	20	0.250	0.750	0.188
5–8	1	4	0	14	19	0.263	0.737	0.194
8–13	0	4	1	12	17	0.235	0.765	0.180
13–14	0	4	1	11	16	0.250	0.750	0.188
14–15	0	4	1	10	15	0.267	0.733	0.196
15–16	0	4	1	9	14	0.286	0.714	0.204
16–18	1	3	0	9	13	0.308	0.692	0.213
18–22	0	2	1	8	11	0.182	0.818	0.149
22–23	1	1	0	8	10	0.200	0.800	0.160
23–25	1	0	0	8	9	0.111	0.889	0.099
25–29	0	0	1	5	6	0.000	1.000	0.000
29–30	0	0	1	4	5	0.000	1.000	0.000
30–31	0	0	1	3	4	0.000	1.000	0.000
31–37	0	0	1	2	3	0.000	1.000	0.000
37–80	0	0	1	1	1	0.000	1.000	0.000
Total	5	—	12	—	—	2.898	14.102	2.166

The third table, $t_3 = 5$ (time interval 4 to 5 months), is

$t_3 = 5$	Dead	Alive	Total
Smoker	0	5	5
Nonsmoker	1	14	15
Total	1	19	20

The individual censored after 2 months (2$^+$) is included in the first table but is excluded from consideration in subsequent tables because his survival time does not span the entire second interval. Table 7.1 displays the complete sequence of 17 tables (strata) for the smoking/survival data.

After accounting for the censored observations (Table 7.1), a log-rank procedure is not different in principle from the analysis of the independence between two binary variables from series of 2×2 tables in general (sometimes referred to as the *Cochran–Mantel–Haenszel test for independence*). For each table (stratum), an estimated number of deaths is calculated as if the risk factor and survival outcome were unrelated (null hypothesis). The expected number of deaths in the ith table when the risk factor is present (F) is then

$$\text{estimated number of deaths } (F): A_i = \left[\frac{a_i + b_i}{n_i} \right] (a_i + c_i)$$

and the corresponding observed number is a_i. The observed/expected difference $a_i - A_i$ reflects either random variation or both random variation and a systematic influence from the risk factor. The comparison is not influenced by the survival time, because both a_i and A_i are calculated for the same stratum (the same survival time, t_i). As with the product-limit estimate of a survival probability, the strata-specific comparisons are also not biased by censored observations. Therefore, a summary estimate constructed from these unbiased estimates is also unbiased.

The term "expected" has a technical meaning. In a statistical context, this term refers to a theoretical value or values calculated as if specified conditions are exactly fulfilled. Thus, expected values are treated as fixed values (not subject to sampling variation). The value represented by A_i (the expected number of deaths among smokers at time t_i) is an example of such an expected value.

The variance of the distribution of the a_i-counts (smokers who died) is estimated with the expression

$$\text{variance}(a_i) = \hat{v}_i = \frac{(a_i + b_i)(a_i + c_i)(c_i + d_i)(b_i + d_i)}{n_i^2(n_i - 1)}$$

for each table (ith table). When $a_i + b_i = 1$, the case when no identical survival times occur, the expression for the same variance estimate is

$$\text{variance}(a_i) = \hat{v}_i = \frac{(a_i + b_i)(c_i + d_i)}{n_i^2}.$$

This estimated variance is also encountered as part of a specific kind of analysis of a 2×2 table called *Fisher's exact test*. A detailed description of this test, used to assess association in a single 2×2 table, is found in many first-year statistics texts.

Each complete survival time generates a table. Each table generates three values; an observed value (a_i), a theoretical value (A_i), and an estimated variance $(\hat{v}_i)$. For the first 2×2 table (0 to 1 month), one death $(a_1 = 1)$ occurred among smokers, $A_1 = 6(1)/23 = 0.261$ deaths are expected when smoking and survival are unrelated, and the estimated variance of the count a_1 is $\hat{v}_1 = 6(17)/23^2 = 0.193$. All 17 stratum-specific values of a_i, A_i, and $\hat{v}_i$ are given in Table 7.1.

Statistics summarizing the association between risk factor and survival time for the entire sequence of 2×2 tables are as follows: (1) the total number of deaths among individuals with the risk factor, represented by $\sum a_i$, (2) the total number of deaths among individuals with the risk factor estimated as if the risk factor and survival status were unrelated, represented by $\sum A_i$, and (3) the variance of the summary $\sum a_i$, represented by $\sum \hat{v}_i$. For the African-American AIDS data, these sums are found in the last row of Table 7.1.

The comparison of $\sum a_i$ and $\sum A_i$ indicates the overall risk/survival association. A formal test statistic measuring the overall strength of the association is

$$X^2 = \left[\frac{\sum(a_i - A_i)}{\sqrt{\text{variance}\left(\sum a_i\right)}} \right]^2 = \frac{\left[\sum(a_i - A_i)\right]^2}{\text{variance}\left(\sum a_i\right)} = \frac{\left[\sum a_i - \sum A_i\right]^2}{\sum \hat{v}_i},$$

where $\sum(a_i - A_i)$ is assumed to have at least an approximate normal distribution. The test statistic X^2 then has an approximate chi-square distribution with one degree of freedom when $\sum a_i$ and $\sum A_i$ differ by chance alone (no association). As with most single summary measures of association, the chi-square statistic measures the risk/survival association only when each deviation $a_i - A_i$ reflects the same underlying value within each stratum (homogeneous).

From the SFMHS data in Table 7.1, these summary values are

$$\sum a_i = 5, \quad \sum A_i = 2.898, \quad \text{and} \quad \sum \hat{v}_i = 2.166.$$

A greater number of deaths exist among smokers (F) relative to the number estimated as if no association existed (5 versus 2.898). However, the chi-square test statistic

$$X^2 = \frac{(5 - 2.898)^2}{2.166} = 2.040$$

generates a p-value of $P(X^2 \geq 2.040 \mid$ risk and survival status unrelated$) = 0.153$, indicating that this increased number of deaths is plausibly due to random sampling variation.

Essentially the same chi-square procedure is frequently applied to any series of 2×2 tables. However, when the sequence of tables is generated from censored survival data, the procedure is called a *log-rank test*. To this point, the description of the log-rank test refers to neither logarithms nor ranks. These terms come from an alternative derivation of the same technique [3].

The application of a chi-square distribution is not strictly correct. The estimated variance of a sum [variance$(\sum a_i)$] is the sum of the estimated variances [$\sum$ variance(a_i)] only when the values a_i are uncorrelated (Chapter 4). This is not the case among the 2×2 log-rank tables. All but the first table contain participants from earlier tables, introducing a table-to-table association. This lack of independence, however, has only a minor influence on the accuracy of the summary chi-square test statistic.

It is important to remember that the log-rank procedure requires that censoring be random (noninformative). A number of techniques are available to study this issue. Some examples are basic statistical comparisons (censored versus complete), descriptive plots, and cluster and logistic regression techniques. In all cases, the goal is to identify the possibility that the censoring is related to survival time. The number of censored observations is not nearly as important as the reason for the censoring. In addition, it is implicitly assumed that censoring and group membership are unrelated. In light of these two requirements (typically assumptions), it is always a good idea to examine the censored observations for nonrandom patterns. For example, the number of randomly censored observations in each group should be approximately proportional to the size of the group.

An alternative approach to assessing an association between a risk factor and survival measured across a sequence of time intervals (strata) consists of contrasting the difference in the number of deaths among the individuals with and without the risk factor. Evidence of a nonrandom difference indicates a likely risk factor influence on survival time. Analogously to the log-rank test, the total numbers of deaths among the individuals with ($\sum a_i$) and without the risk factor ($\sum c_i$) are counted. The corresponding number of expected deaths among individuals with ($\sum A_i$) and without ($\sum C_i$) the risk factor are estimated again as if risk factor and survival time were unrelated. For the

ith-interval (table), the number of observed deaths is c_i and the theoretical number of deaths C_i is

$$\text{estimated number of deaths } (\bar{F})\text{: } C_i = \left[\frac{a_i + c_i}{n_i}\right](c_i + d_i)$$

for individuals without the risk factor (column eight in Table 7.1). When all survival times are unique (no ties), $c_i = 1 - a_i$ and $C_i = 1 - A_i$.

The two summary values ($\sum c_i$ and $\sum C_i$) and the two previously calculated summary values ($\sum a_i$ and $\sum A_i$) form the natural chi-square test statistic:

$$X^2 = \sum \frac{(\text{observed}_i - \text{expected}_i)^2}{\text{expected}_i} = \frac{\left(\sum a_i - \sum A_i\right)^2}{\sum A_i} + \frac{\left(\sum c_i - \sum C_i\right)^2}{\sum C_i}.$$

The value X^2 has an approximate chi-square distribution with one degree of freedom when risk factor and survival status are unrelated. Again for the SFMHS data, the differences in summary values $\sum a_i = 5$ versus $\sum A_i = 2.898$ among smokers and $\sum c_i = 12$ versus $\sum C_i = 14.102$ among non-smokers (Table 7.1) are evaluated with the chi-square test statistic

$$X^2 = \frac{(5 - 2.898)^2}{2.898} + \frac{(12 - 14.102)^2}{14.102} = 1.838,$$

yielding a p-value of $P(X^2 = 1.838 \mid \text{risk and survival time unrelated}) = 0.175$.

The test statistic X^2 from this version of a chi-square test is always less than the log-rank chi-square value calculated from the same data, but both address the identical issue. A feature of this alternative approach is the possibility of identifying an association between risk and survival time in a sequence of $2 \times k$ tables (k-level risk factor).

Two-sample analysis: exponential proportional hazards model

When no censoring occurs, it is possible to represent survival time (t) and its relationship to a series of explanatory variables (x) with a linear regression model (such as $t_i = a + b_1 x_{i1} + b_2 x_{i2} + \cdots$). However, survival data without censored observations are relatively rare. Of more importance, estimation of survival probabilities and hazard functions is key to understanding the underlying mechanisms of the risk/survival relationship. The relevance of survival probabilities and hazard functions to the description of survival data has been discussed and becomes even clearer in the following. In addition, unlike data typically analyzed with linear regression models, survival times are rarely normally distributed and are more accurately described by right-skewed distributions.

An effective two-sample analysis based on a parametric model begins with the comparison of two hazard functions, one from each population sampled, denoted $h_0(t)$ and $h_1(t)$. The difference between these two hazard functions is measured by a constant ratio or

$$\frac{h_1(t)}{h_0(t)} = c \quad \text{or} \quad h_1(t) = h_0(t) \times c.$$

That is, the two-sample hazards model requires the hazard rates from each sampled population to be proportional for all survival times. When c is greater than one, $h_1(t) > h_0(t)$, and when c is less than one, $h_1(t) < h_0(t)$. An obvious and important case occurs when $c = 1$, making $h_0(t) = h_1(t)$. Nevertheless, the ratio is the same for all survival times. For the exponential distribution of survival times, the hazard function is constant (does not depend on time). Therefore, the two exponential model hazard functions $h_0(t)$ and $h_1(t)$ are necessarily proportional.

When the comparison of proportional hazard functions is expressed in terms of two parameters, the two-sample exponential model becomes

$$h(t \mid F) = e^{b_0 + b_1 F},$$

where $F = 0$ produces a hazard rate called the *baseline hazard rate* and $F = 1$ produces a second and proportional hazard rate. Specifically, the two-sample hazards model consisting of parameters b_0 and b_1 becomes

$$F = 0: h_0(t) = e^{b_0} \quad \text{and} \quad F = 1: h_1(t) = e^{b_0 + b_1} = h_0(t)e^{b_1},$$

making the ratio of the hazard functions $c = h_1(t)/h_0(t) = e^{b_1}$. The model is completely determined by two parameters, b_0 and b_1. Furthermore, the parameter b_1 becomes the primary summary measure of the difference between the two samples of survival data. For example, when no difference exists between the survival times from two compared groups, $b_1 = 0$, then $c = 1$ and $h_1(t) = h_0(t)$.

The proportional hazards model is based on the expression $e^{b_0 + b_1 F}$ for three reasons. It requires the hazard function associated with one of the sampled populations to be a product of two components, a baseline hazard function and a multiplicative constant (proportional). Thus, the comparison between groups results in a familiar form similar to the relative risk or odds ratio measure of association. It is essentially a rate ratio. In addition, the hazard ratio model naturally generalizes (as will be seen) to other survival distributions and is easily extended to incorporate any number of explanatory variables. And, the estimate of the hazard ratio c is positive under all circumstances.

For the SFMHS data, the white participants ($n = 174$) consisted of 80 men who did not smoke ($n_0 = 80$ with $d_0 = 71$ deaths and $n_0 - d_0 = 9$ censored observations) and 94 men who did smoke ($n_1 = 94$ with $d_1 = 84$ deaths and $n_1 - d_1 = 10$ censored observations). These data illustrate the application of the exponential two-sample proportional hazards model. Although the comparison of two samples from exponential survival time distributions is simple and direct (comparison of two mean values), a fully developed analysis based on a proportional hazards model serves to introduce the more complicated analyses using other survival distributions applied to more extensive multivariable data.

As with all statistical models, the collected data are used to estimate the model parameters. Although the parameters of a survival model typically require computer estimation, the maximum likelihood estimates for the exponential model parameters can be calculated directly. Again, the estimated mean survival times are the key. For each SFMHS group, they are as follows:

nonsmokers: estimated mean survival time $= \hat{\mu}_0 = \sum t_i / d_0 = 24.775$

months, making $\hat{h}_0(t) = \hat{\lambda}_0 = \dfrac{1}{\hat{\mu}_0} = \dfrac{1}{24.775} = 0.040$ and

smokers: estimated mean survival time $= \hat{\mu}_1 = \sum t_i / d_1 = 21.655$

months, making $\hat{h}_1(t) = \hat{\lambda}_1 = \dfrac{1}{\hat{\mu}_1} = \dfrac{1}{21.655} = 0.046.$

Table 7.2. Exponential proportional hazards model parameter estimates from the SFMHS—nonsmokers ($n_0 = 80$) versus smokers ($n_1 = 94$).

Variables	Symbols	Estimates	Std. errors	p-values
Baseline	b_0	−3.210	—	—
Nonsmoker/smoker	b_1	0.135	0.161	0.404
	LogLikelihood $= -641.216$			

The estimated exponential hazards model parameters are then $\hat{b}_0 = \log(\hat{\lambda}_0) = \log(0.040) = -3.210$ and $\hat{b}_1 = \log(\hat{\lambda}_1/\hat{\lambda}_0) = \log(\hat{\mu}_0/\hat{\mu}_1) = \log(24.775/21.665) = 0.135$. The estimated proportional hazards model becomes

$$\hat{h}(t \mid F) = e^{\hat{b}_0 + \hat{b}_1 F} = e^{-3.210 + 0.135F}.$$

More specifically,

nonsmokers ($F = 0$): $\hat{h}_0(t) = \hat{\lambda}_0 = e^{\hat{b}_0} = e^{-3.210} = 0.040$

baseline hazard function

and

smokers ($F = 1$): $\hat{h}_1(t) = \hat{\lambda}_1 = e^{\hat{b}_0 + \hat{b}_1} = e^{-3.210 + 0.135}$

$$= (0.040)(1.144) = 0.046.$$

The estimated constant ratio of hazard functions is $\hat{c} = h_1(t)/h_0(t) = e^{\hat{b}_1} = e^{0.135} = 1.144$ (also denoted hr) and $hr = 0.046/0.040 = 1.144$. The estimates $\hat{b}_0$ and $\hat{b}_1$ are maximum likelihood estimates because they are functions of the maximum likelihood estimated mean values $\hat{\mu}_0$ and $\hat{\mu}_1$. The estimated hazards model parameters are summarized in Table 7.2.

Survival functions based on the estimated hazard rates $\hat{\lambda}_0 = 0.040$ and $\hat{\lambda}_1 = 0.046$ are directly estimated (also maximum likelihood estimates). These estimates are

$$\hat{S}_0(t) = e^{-\hat{\lambda}_0 t} = e^{-0.040t}   \text{for nonsmokers}   \text{and}$$
$$\hat{S}_1(t) = e^{-\hat{\lambda}_1 t} = e^{-0.046t}   \text{for smokers.}$$

The estimated survival curve $\hat{S}_0(t)$ is always above the estimated survival curve $\hat{S}_1(t)$ because $\hat{b}_1$ is greater than zero, which only occurs when $\hat{\lambda}_1$ is greater than $\hat{\lambda}_0$, making $S_0(t) > S_1(t)$ for any time t.

Once the model parameters are estimated, they provide a variety of ways to describe the differences in survival experience. For example, the reciprocal relationship between mean survival time and risk measured by a rate is once again apparent. That is,

$$\text{mean survival time} = \frac{1}{\text{hazard function}} = \frac{1}{h(t \mid F)} = \frac{1}{e^{b_0 + b_1 F}}.$$

For the SFMHS data, for nonsmokers

$$\text{mean survival time } (F = 0) = \frac{1}{\hat{h}(t \mid F = 0)} = \frac{1}{e^{b_0}} = \frac{1}{e^{-3.210}} = 24.775$$

and for smokers

$$\text{mean survival time } (F = 1) = \frac{1}{\hat{h}(t \mid F = 1)} = \frac{1}{e^{b_0 + b_1}} = \frac{1}{e^{-3.210 + 0.135}}$$

$$= 21.655.$$

The model parameters produce estimates of the median survival time for each sampled population, given by the expression

$$\hat{t}_F = \log(2)e^{-(\hat{b}_0 + \hat{b}_1 F)} = \log(2)e^{-(-3.210 + 0.135 F)}.$$

Thus, for nonsmokers, the estimated median survival time is $\hat{t}_0 = \log(2)(24.775) = 17.172\,(F = 0)$ and, for smokers, the estimated median survival time is $\hat{t}_1 = 0.693\,(21.655) = 15.010\,(F = 1)$. Furthermore, the ratios of the model-estimated hazard rates, mean values, and median values are identical. From the exponential model, the ratios are $\hat{c} = \hat{\lambda}_1/\hat{\lambda}_0 = \hat{\mu}_0/\hat{\mu}_1 = \hat{t}_0/\hat{t}_1$ in general and, specifically, for the AIDS data, they are $\hat{c} = 1.144$.

The question certainly arises: Is the observed difference between sample estimates due to chance (for the example, $\hat{\lambda}_0 = 0.040$ compared to $\hat{\lambda}_1 = 0.046$ or $\hat{t}_0 = 17.172$ compared to $\hat{t}_1 = 15.010$ or $\hat{c} = 1.144$ compared to 1.0) or does evidence exist of a systematic difference between compared groups? In symbols: Is $b_1 = 0$ or $b_1 \neq 0$? Three essentially equivalent statistical approaches help provide an answer. They are as follows:

1. A normal-distribution-based statistical test: the maximum likelihood estimate $\hat{b}_1$ has an approximate normal distribution with estimated variance of 0.026 (Table 7.2), yielding the test statistic

$$z = \frac{\hat{b}_1 - 0}{\sqrt{\text{variance}(\hat{b}_1)}} = \frac{0.135 - 0}{\sqrt{0.026}} = 0.835.$$

The value z has an approximate standard normal distribution when no difference exists between the compared groups ($b_1 = 0$ or $\mu_0 = \mu_1$ or $\lambda_0 = \lambda_1$ or $c = 1$). The corresponding p-value of $P(|Z| \geq 0.835 \mid b_1 = 0) = 0.404$ provides little evidence of a systematic difference between the survival patterns of nonsmokers and smokers. As noted, such a test of a maximum likelihood estimated parameter is frequently called *Wald's test* (z^2).

For the exponential hazards model, the test of the estimated coefficient $\hat{b}_1$ is identical to a comparison of the log-mean survival times or

$$z = \frac{\hat{b}_1 - 0}{\sqrt{\text{variance}(\hat{b}_1)}} = \frac{[\log(\hat{\mu}_0) - \log(\hat{\mu}_1)] - 0}{\sqrt{\dfrac{1}{d_0} + \dfrac{1}{d_1}}}$$

$$= \frac{\log(24.775) - \log(21.665) - 0}{\sqrt{\dfrac{1}{71} + \dfrac{1}{84}}} = 0.835.$$

2. A normal-distribution-based confidence interval: an approximate 95% confidence interval based on the estimated parameter $\hat{b}_1$ from the SFMHS smoking data and the normal distribution is

$$\hat{b}_1 \pm 1.960\sqrt{\text{variance}(\hat{b}_1)} = 0.135 \pm 1.960\sqrt{0.026} \text{ or } (-0.181, 0.451).$$

As usual, the probability is approximately 0.95 that the underlying model parameter b_1 is contained in the estimated confidence interval $(-0.181, 0.451)$. The value zero is contained in the interval. The value zero is thus a plausible value for the parameter b_1, again providing no persuasive evidence of a systematic influence from smoking exposure. In other words, the conjecture that the parameter $b_1 = 0$ appears consistent with the range of likely parameter values -0.181 to 0.451 in light of the variation observed in the sampled data. Alternatively, the approximate 95% confidence interval based on the estimated hazard ratio $\hat{c} = \hat{hr} = e^{0.135} = 1.144$ has a lower bound $= e^{-0.181} = 0.835$ and an upper bound $= e^{0.451} = 1.569$ and contains the value one ($c = hr = 1.0$).

3. A likelihood ratio comparison: contrasting two log-likelihood values effectively identifies systematic differences between two nested survival time models. First, a log-likelihood value is estimated under the condition that $b_1 = 0$ ($h_0(t) = h_1(t)$ or $c = 1$) and then under the condition that $b_1 \neq 0$ ($h_0(t) \neq h_1(t)$ or $c \neq 1$). The comparison of the log-likelihood values likely reflects any important difference in survival time between the compared groups (significant?). For the AIDS smoking data, the two relevant log-likelihood values from the exponential hazards models are

no two-sample difference exists ($b_1 = 0$): $\log(L_{b_1=0}) = -641.566$

and

a two-sample difference exists ($b_1 \neq 0$): $\log(L_{b_1 \neq 0}) = -641.216$.

The likelihood ratio test-statistic

$$X^2 = -2[\log(L_{b_1=0}) - \log(L_{b_1 \neq 0})]$$
$$= -2[-641.566 - (-641.216)] = 0.699$$

has an approximate chi-square distribution with one degree of freedom when $b_1 = 0$. The p-value is $P(X^2 \geq 0.699 \mid b_1 = 0) = 0.403$. The three approaches essentially agree (note: $X^2 = 0.699 \approx (0.835)^2 = z^2$). Furthermore, these three approaches to a two-sample comparison generally give similar results.

Aside. A confidence interval for a specific parameter (denoted g) is related to a Wald test of the hypothesis that the estimated value (denoted $\hat{g}$) differs from zero by chance alone. Specifically, when the value zero is not contained in a 95% confidence interval, the parallel test of the hypothesis $g = 0$ produces a p-value less than 0.05.

The reason these two approaches produce the same answer is as follows: For a 95% confidence interval based on the normal distribution of the estimate $\hat{g}$, when either the lower bound $= \hat{g} - 1.960 S_{\hat{g}} > 0$ or the upper bound $= \hat{g} + 1.960 S_{\hat{g}} < 0$ (zero excluded), the Wald test statistic $\hat{g}/S_{\hat{g}}$ is necessarily either greater than 1.960 or less than 1.960, causing the significance perbability to be less than 0.05.

A visual "analysis" provides another perspective on the differences between estimated hazard functions. The interpretation of a graphic comparison is

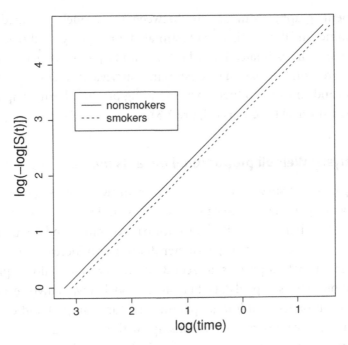

Figure 7.2. "Log–log" transformed survival functions using the SFMHS data–nonsmokers ($n_0 = 80$) versus smokers ($n_1 = 94$).

certainly more subjective than a statistical test but yields a direct and intuitive impression of the two-sample difference. As previously described, a "log–log" transformation of an exponential survival function produces a straight line with intercept $\log(\lambda)$ and slope 1.0 when plotted for values $\log(t)$. The plots of the estimated lines $\hat{l}_0(t) = \log(-\log[\hat{S}_0(t)]) = \log(\hat{\lambda}_0) + \log(t_i) = -3.210 + \log(t_i)$ for nonsmokers and $\hat{l}_1(t) = \log(-\log[\hat{S}_1(t)]) = \log(\hat{\lambda}_1) + \log(t_i) = -3.075 + \log(t_i)$ for smokers are displayed in Figure 7.2. The vertical distance between the two parallel lines (slopes $= 1$) is $\hat{l}_1(t) - \hat{l}_0(t) = \log(\hat{\lambda}_1) - \log(\hat{\lambda}_0) = \log(\hat{\lambda}_1/\hat{\lambda}_0) = \hat{b}_1$, directly displaying the logarithm of the estimated ratio of the hazard functions $h_1(t)$ and $h_0(t)$. For the smoking data, this distance is $\hat{b}_1 = 0.135$ ($\hat{hr} = e^{\hat{b}_1} = 1.144$). Visually, these two lines show almost no separation. Again, the difference in survival times between nonsmokers and smokers at best appears slight.

Formal statistical methods exist to evaluate the correspondence between an exponential model and the sampled data. These goodness-of-fit techniques are discussed in Chapters 8 and 9. In addition to a formal statistical assessment, graphic comparisons between the model-generated survival probabilities and the product-limit estimated survival probabilities are always a good idea. The estimated hazard rates $\hat{\lambda}_0$ and $\hat{\lambda}_1$ allow visual comparisons of the corresponding parametric estimated survival functions for both non-smokers and smokers to their corresponding product-limit nonparametric estimated survival functions (Figure 7.3).

Two-sample analysis: Weibull proportional hazards model

A comparison of survival data collected from two groups can be based on a number of parametric descriptions of the sampled populations. The two-parameter Weibull probability distribution is a common and frequently effective choice. Although a two-parameter distribution increases the complexity of the estimation process, a second parameter potentially improves the description of the sampled data. Statistical models always involve a trade-off between complexity (additional parameters, for example) and obtaining a clear and simple sense of the relationships within the data.

As with the previous exponential survival model, a constant ratio of two hazard functions is the foundation of the two-sample comparison. The two hazard functions are again postulated to be proportional. In symbols,

$$h_1(t) = h_0(t) \times c.$$

However, the Weibull model hazard function is not constant, making the following description a bit more complex than the exponential case but not different in principle.

The Weibull two-sample proportional hazards model becomes

$$h(t \mid F) = e^{(b_0 + b_1 F)\gamma} \gamma t^{\gamma - 1} = \lambda_F \gamma t^{\gamma - 1}$$

and

when $F = 0$: $h_0(t) = e^{b_0 \gamma} \gamma t^{\gamma - 1} = \lambda_0 \gamma t^{\gamma - 1}$ baseline hazard function,

where the scale parameter λ_0 equals $e^{b_0 \gamma}$ and

when $F = 1$: $h_1(t) = e^{(b_0 + b_1)\gamma} \gamma t^{\gamma - 1} = \lambda_0 \gamma t^{\gamma - 1} e^{b_1 \gamma} = h_0(t) e^{b_1 \gamma}$,

where the scale parameter λ_1 equals $e^{(b_0 + b_1)\gamma}$. Thus, the constant ratio of the two hazard functions is $c = hr = e^{b_1 \gamma}$. As in the exponential model, the analytic focus is again on the model parameter b_1.

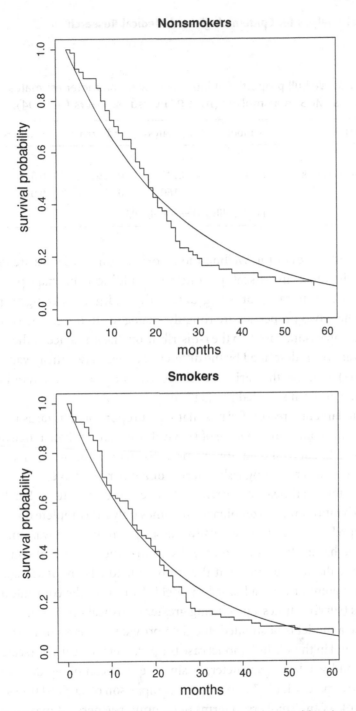

Figure 7.3. Comparison of parametric estimated exponential survival functions for nonsmokers and smokers (continuous line) to their product-limit nonparametric estimates (step function).

Table 7.3. Weibull proportional hazards model parameter estimates from the SFMHS–nonsmokers ($n_0 = 80$) versus smokers ($n_1 = 94$).

Variables	Symbols	Estimates	Std. errors	p-values
Baseline	b_0	−3.240	—	—
Nonsmoker/smoker	b_1	0.135	0.139	0.331
Shape	γ	1.160	0.072	0.017
	LogLikelihood $= -638.605$			

The influences of the Weibull model explanatory variables are described by the differences in the scale parameters. In addition, the shape parameter is required to be the same for both groups compared, guaranteeing proportionality. These two properties are roughly analogous to the classic two-sample t-test comparison. Based on the t-distribution, the influence of the "explanatory" variable is described by the difference between two mean values (scale: $\mu_1 - \mu_2$), whereas the variances (shape: $\sigma_1^2 = \sigma_2^2 = \sigma^2$) are required to be the same for both sampled populations.

A fundamental reason for postulating a proportional hazards model is to capture the continuous nature of survival time data. When a logistic model is applied to survival data, for example, the natural measure of association is an odds ratio reflecting risk in terms of a binary variable, which does not account for the time of occurrence of the event under study. Furthermore, the odds ratio summarizes binary outcomes only for complete observations. A comprehensive and more sensitive description of a survival time pattern based on hazard functions emerges by comparing values that continuously vary over the time period that the study individuals are observed. From a practical point of view, a hazards model fully utilizes the continuous observations (survival times), producing greater statistical precision.

Once a model is postulated, the data provide estimates of the parameter values, and in the Weibull model case the parameters are b_0, b_1 (scale parameters), and γ (shape parameter). Using the AIDS smoking data again, the model estimates (Table 7.3) produce a comparison of survival times between nonsmokers and smokers in terms of computer-generated maximum likelihood estimated parameters, particularly the model coefficient b_1.

The estimated shape parameter is $\hat{\gamma} = 1.160$ (Table 7.3) but the p-value is calculated based on the logarithm of the parameter. As with most bounded parameters $(\gamma > 0)$, the distribution of the logarithm of the estimate $\hat{\gamma}$ produces a more symmetric ("normal-like") distribution. That is, the test statistic

$$z = \frac{\log(\hat{\gamma}) - \log(1.0)}{\sqrt{\text{variance}[\log(\hat{\gamma})]}} = \frac{\log(1.160) - 0}{\sqrt{0.00389}} = 2.379$$

is a single observation from an approximate standard normal distribution when $\gamma = 1.0$ (constant hazard ratio), yielding a p-value of $P(|Z| \geq 2.379 \mid \gamma = 1.0) = 0.017$. Thus, the Weibull hazards model $(\gamma \neq 1)$ likely provides a substantially better description of the SFMHS smoking exposure data than the exponential hazards model $(\gamma = 1)$.

The Weibull proportional hazards model based on estimated parameters from the SFMHS smoking data becomes

$$\hat{h}(t \mid F) = e^{(-3.240 + 0.135F)1.160}1.160t^{0.160}.$$

More specifically, the baseline hazard function is

when $F = 0$: $\hat{h}_0(t) = e^{-3.240(1.160)}1.160t^{0.160} = 0.023(1.160)t^{0.160}$

and

when $F = 1$: $\hat{h}_1(t) = \hat{h}_0(t)e^{0.135(1.160)} = \hat{h}_0(t)1.170,$

making the ratio of hazard functions $\hat{h}_1(t)/\hat{h}_0(t) = 1.170$ (or $\hat{c} = \hat{h}r = e^{\hat{b}_1\hat{\gamma}} = e^{0.135(1.160)} = 1.170$). The two estimated hazard functions $\hat{h}_0(t)$ and $\hat{h}_1(t)$ are displayed in Figure 7.4.

A fundamental property of a Weibull distribution survival model is that the hazard rates themselves are not constant but the ratio of the hazard rates is constant. Although it is not visually obvious from Figure 7.4, the ratio of the estimated model hazard functions is the same for all survival times t [$\hat{h}r = \hat{h}_1(t)/\hat{h}_0(t) = e^{\hat{b}_1\hat{\gamma}} = 1.170$]. As required, the model hazard functions are proportional. The critical importance of this property is that a single estimate $\hat{b}_1$ (or $\hat{h}r$) then accurately summarizes the differences between two groups. This single summary becomes the focus of testing, confidence intervals, inferences, and interpretations unaffected by any specific survival time. When a hazard ratio is not constant, the comparison is not interpreted simply in terms of a single parameter.

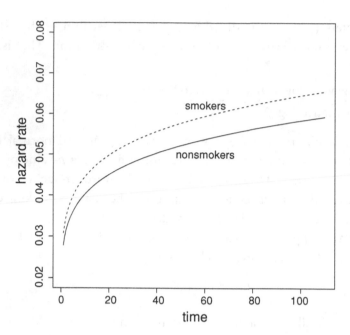

Figure 7.4.   Weibull model hazard functions estimated for nonsmokers and smokers from the AIDS smoking data ($n = 174$).

Using the mean survival time (Chapter 5) produces the parallel two-sample expression

$$\text{mean survival time} = \mu_F = \lambda_F^{-1/\gamma}\Gamma(1 + 1/\gamma), \quad \text{where } \lambda_F = e^{(b_0 + b_1 F)\gamma},$$

and yields an estimated mean survival time for each group ($F = 0$ and $F = 1$). Applied to the AIDS smoking data ($n = 174$), the estimated mean survival times are $\hat{\mu}_0 = 24.245$ months for nonsmokers ($F = 0$) and $\hat{\mu}_1 = 21.180$ months for smokers ($F = 1$). The ratio of these estimates is, like the hazard ratio, constant and is $\hat{\mu}_0/\hat{\mu}_1 = e^{\hat{b}_1} = 1.145$.

The median survival time t_m is

$$S(t_m) = \tfrac{1}{2} = e^{-\lambda_F t_m^{\gamma}} = e^{-\exp[(b_0 + b_1 F)\gamma]t_m^{\gamma}},$$

giving a model-estimated median value for each group as

$$\hat{t}_m = [\log(2)]^{1/\hat{\gamma}} e^{-(\hat{b}_0 + \hat{b}_1 F)}.$$

For the two-sample smoking data, the estimated median survival times are

nonsmokers $(F = 0)$: $\hat{t}_0 = [\log(2)]^{1/1.160} e^{3.240} = 18.622$

and

smokers $(F = 0)$: $\hat{t}_0 = [\log(2)]^{1/1.160} e^{3.105} = 16.268$.

Identical to the mean value, the ratio of these model estimated median values is $\hat{t}_0/\hat{t}_1 = e^{\hat{b}_1} = 1.145$ for any survival time t.

The key summary parameter is again the coefficient b_1. For the smoking AIDS data, the Weibull proportional hazards estimate is $\hat{b}_1 = 0.135$ (Table 7.3). Although the estimate of b_1 is essentially equal to the estimate from the exponential model applied to the same data, an increase in the precision is achieved from the two-parameter model [standard errors of $\hat{b}_1$: 0.139 (Weibull) and 0.161 (exponential)].

When $b_1 = 0$, the hazard ratio is $hr = h_1(t)/h_0(t) = 1.0$ (constant hazard rate). Thus, the model coefficient b_1 is a single measure of the difference between the two proportional Weibull hazard functions, much as in the exponential model. As described in the context of the exponential model, three equivalent ways exist to assess the influence of random variation on the estimated parameter $\hat{b}_1$. The comparison of log-likelihood values is one of these methods. Again, the question becomes: Is $b_1 = 0$ or $b_1 \neq 0$? For the smoking data, the specific estimated log-likelihood values are

no two-sample difference exists $(b_1 = 0)$: $\log(L_{b_1=0}) = -639.078$

and

a two-sample difference exists $(b_1 \neq 0)$: $\log(L_{b_1 \neq 0}) = -638.605$,

producing the likelihood ratio chi-square test-statistic

$$X^2 = -2[\log(L_{b_1=0}) - \log(L_{b_1 \neq 0})] = -2[-639.078 - (-638.605)] = 0.945.$$

The associated p-value is $P(X^2 \geq 0.945 \mid b_1 = 0) = 0.331$. A significance probability of 0.331 indicates that the estimate $\hat{b}_1 = 0.135$ and the parameter $b_1 = 0$ plausibly differ by chance alone, providing again no substantial evidence of a systematic influence from smoking exposure on survival time.

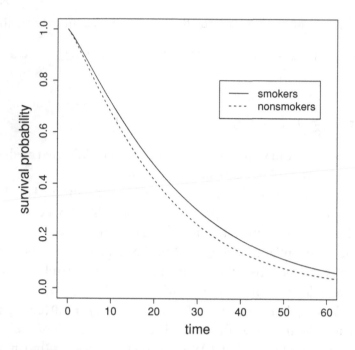

Figure 7.5.   Weibull model estimated survival curves for nonsmokers and smokers–SFMHS data ($n = 174$).

The two estimated Weibull survival functions are

$$\hat{S}_0(t) = e^{-\hat{\lambda}_0 t^{\hat{\gamma}}} = e^{-0.023 t^{1.160}} \qquad \text{nonsmokers}$$

and

$$\hat{S}_1(t) = e^{-\hat{\lambda}_1 t^{\hat{\gamma}}} = e^{-0.027 t^{1.160}} \qquad \text{smokers,}$$

where, as before, the estimated scale parameter is $\hat{\lambda}_F = e^{(\hat{b}_0 + \hat{b}_1 F)\hat{\gamma}} = e^{(-3.240 + 0.135 F)1.160}$ and $F = 0$ (nonsmokers) and $F = 1$ (smokers). A graphic comparison of the estimated survival curves associated with non-smokers and smokers is displayed in Figure 7.5. Visually, these estimates are consistent with the likelihood analysis and show only slight differences between nonsmokers and smokers from the SFMHS data.

When a Weibull proportional hazards model is used to identify differences between two samples of survival data, the shape parameter must be the same for both groups. When the shape parameter γ differs between compared groups, the hazard ratio is not constant (not proportional) and, therefore, not accurately summarized by a single value.

Critical to all models postulated to have proportional hazard functions is the following question: Are the hazard functions proportional? A first step in exploring this question is to divide the data into groups (for the example, nonsmokers and smokers) and plot for each group the log–log transformed survival probabilities from the product-limit estimated survival function (Figure 7.6, top). Also included in Figure 7.6 (bottom) is a plot of two summary least-squares estimated straight lines based on the pairs of values $\log(t_i)$ and $\log(-\log[\hat{P}_i])$ for nonsmokers and smokers. In symbols, the two estimated straight lines are

$$\hat{l}_F(t) = \hat{a}_F + \hat{b}_F \log(t_i),$$

where

$$\hat{l}_0(t) = -3.362 + 1.094 \log(t_i) \qquad \text{nonsmokers} = F = 0 \quad \text{and}$$
$$\hat{l}_1(t) = -3.736 + 1.166 \log(t_i) \qquad \text{smokers} = F = 1.$$

Proportionality requires that these estimated lines be parallel (give or take random variation). This graphic assessment of proportionality is useful when the data analyzed are relatively simple and can be divided into a few meaningful groups. A more extensive data set (for example, one containing several continuous explanatory variables—next two chapters) requires a more sophisticated approach.

For the Weibull two-sample model, the distance between these two parallel lines is the logarithm of the hazard ratio (namely, b_1) or

$$l_1(t) - l_0(t) = [\log(\lambda_1) + \gamma \log(t)] - [\log(\lambda_0) + \gamma \log(t)]$$
$$= \log\left(\frac{\lambda_1}{\lambda_0}\right) = b_1.$$

This expression shows that the shape parameters must be equal ($\gamma_1 = \gamma_0 = \gamma$) or the two Weibull hazard functions are not proportional. That is, the hazard ratio is constant (does not depend on survival time t) only when $\gamma_1 = \gamma_0 = \gamma$, causing the lines denoted $l_0(t)$ and $l_1(t)$ to be parallel and differences are accurately summarized by a single value, namely b_1.

More formally, log-likelihood values provide a straightforward and general method for statistically evaluating the observed differences between two

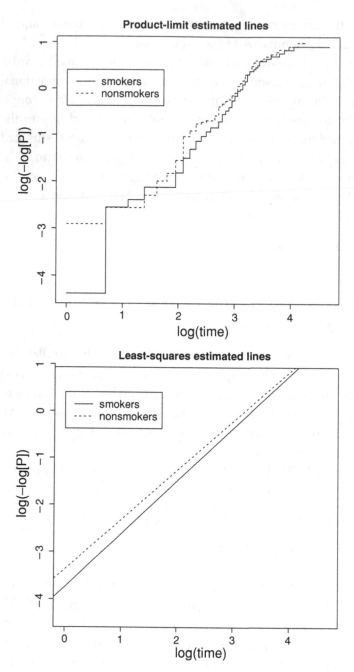

Figure 7.6.   Plots of the log–log transformation product-limit survival probabilities (log[–log(*P*)]) for values of the logarithm of the survival time log(*t*)–nonsmokers versus smokers.

Table 7.4. Summary values for the assessment of the assumption of equal shape parameters (proportionality—$\gamma_0 = \gamma_1 = \gamma$).

	n	$\hat{\lambda}_0$	$\hat{\lambda}_1$	$\hat{\gamma}$	$\log(L)$
Nonsmoker ($\hat{\gamma}_0$)	80	0.024	—	1.151	−297.794
Smoker ($\hat{\gamma}_1$)	94	—	0.027	1.168	−340.805
Combined ($\hat{\gamma}$)	174	0.023	0.027	1.160	−638.605

Weibull shape parameters estimated from two samples, such as the nonsmoking/smoking data from the SFMHS AIDS patients. First, the log-likelihood values are calculated separately from hazards models applied to each group ($\gamma_0 \neq \gamma_1$—not proportional). Then, for a combined analysis, a single log-likelihood value is calculated ($\gamma_0 = \gamma_1 = \gamma$—proportional). Combined, in this context, means that the model accounts for differences between groups (differences in scale) but is based on a single shape parameter common to both groups, namely γ. The sum of the two log-likelihood values (model: $\gamma_0 \neq \gamma_1$) compared to the single log-likelihood value (model: $\gamma_0 = \gamma_1$) rigorously measures the differences between requiring two shape parameters rather than one.

For the AIDS smoking exposure data, the estimated shape parameters and the associated log-likelihood values are given in Table 7.4. The resulting chi-square test statistic is $X^2 = -2[-638.605 - (297.794 + 340.805)] = 0.013$. The observed X^2-value has an approximate chi-square distribution with one degree of freedom when the underlying Weibull hazard functions are proportional. The obvious similarity in the likelihood values (638.605 versus 638.599) is reflected by a p-value of $P(X^2 \geq 0.013 \mid \gamma_0 = \gamma_1 = \gamma) = 0.908$. No evidence exists to suspect the assumption that the Weibull hazard functions from the two smoking exposure groups are proportional (same shape parameter). Simply stated, the estimated shape parameters $\hat{\gamma}_1 = 1.168$ and $\hat{\gamma}_0 = 1.151$ likely differ by chance alone.

One last property of the Weibull distribution two-sample comparison enriches the interpretation of the two-sample proportional hazards model. The ratio of the hazard rates (multiplicative comparison) is

$$\text{hazard ratio} = hr = \frac{h_1(t)}{h_0(t)} = c = e^{b_1 \gamma}$$

and as estimated from the AIDS data becomes $\hat{hr} = 1.170$. In addition, the ratio of the mean and median values is

$$\frac{\mu_0}{\mu_1} = \frac{t_0}{t_1} = e^{b_1}$$

and as estimated from the AIDS data becomes $e^{0.135} = 1.145$. Algebraically, the relationship of the ratio of median values and ratio of mean values to the ratio of hazard rates is

$$hr = \left[\frac{t_0}{t_1}\right]^\gamma = \left[\frac{\mu_0}{\mu_1}\right]^\gamma = \frac{h_1(t)}{h_0(t)} = c$$

and as estimated from the AIDS smoking data becomes $\hat{hr} = 1.145^{1.160} = 1.710$.

A proportional hazards survival model dictates that the measure of a two-sample difference be a multiplicative and constant ratio of hazard functions (hazard ratio = $\hat{hr} = 1.710$). Alternatively, the differences between two groups can be equally described in terms of ratios of mean or median values estimated from a Weibull hazards model, because the ratios $\hat{\mu}_0/\hat{\mu} = \hat{t}_0/\hat{t}_1 = e^{b_1}$. For the smoking data, in terms of mean or median values, their ratios are $\hat{\mu}_0/\hat{\mu} = \hat{t}_0/\hat{t}_1 = e^{0.135} = 1.145$. It is said that nonsmoking "accelerates" the survival time by a factor of 1.145 relative to smokers ($\hat{\mu}_0 = 1.145\hat{\mu}_1$ or $\hat{t}_0 = 1.145\hat{t}_1$). Other *accelerated failure time models* exist generating time scale comparisons, which would be a topic in a more extensive presentation of survival analysis methods. The Weibull proportional hazards model is the only survival model that can be interpreted using either scale (ratio or time scales) [3].

General hazards model: parametric

Introduction

The exponential and Weibull hazards models can be extended naturally from the two-sample case to include any number of explanatory variables. In a model context, explanatory variables have a variety of names. They are called *independent variables* or *predictor variables* or *risk variables* or sometimes just *co-variables*. These variables can be of any kind. They can be binary variables (such as the two-sample model), counts, categorical indicator variables, or continuous measures. Technically, it is said that the explanatory variables are *unrestricted*. The outcome variable (sometimes called the *dependent variable* or *response variable* or *y-variable*) for a survival model is the observed survival time and continues to be denoted by t.

As with the two-sample model, a convenient way to assess the influence of a set of explanatory variables on survival time is a comparison of proportional hazard functions. Once again, the form of the proportional hazards model is

$$h_i(t) = h_0(t) \times c_i.$$

To incorporate the influences of k explanatory variables (denoted x_1, $x_2, \ldots, x_k$) into a survival model, a constant of proportionality c_i is constructed that is a function of the dependent variables. Specifically, the model becomes

$$
\begin{aligned}
h_i(t \mid x_{i1}, x_{i2}, \ldots x_{ik}) &= h_0(t) \times c_i \\
&= h_0(t) \times e^{b_1 x_{i1} + b_2 x_{i2} + \cdots + b_k x_{ik}} \\
&= h_0(t) \times e^{\sum b_j x_{ij}}.
\end{aligned}
$$

Thus, the hazard ratio is $c_i = hr_i = e^{\sum b_j x_{ij}}$, where $j = 1, 2, \ldots, k =$ number of explanatory variables. This expression describes an *additive proportional hazards model* relating the hazard function $h_i(t)$ for the *i*th person, observation, or group to a baseline hazard function $h_0(t)$ for a specific set of explanatory variables. For a single binary explanatory variable $x_1 = F$, the multivariable model reduces to the previous two-sample case, where

$$h(t \mid F) = h_0(t)e^{bF} = h_0(t) \times c \quad \text{and} \quad c = hr = e^{bF}.$$

When the baseline hazard function is a specific parametric function (this chapter), the model is naturally called *parametric.* The parametric approach, as might be expected, requires a complete description of the specific probability distribution that produced the sampled survival times. It is, however, not always necessary to define explicitly the sampled probability distribution. The model is then called *semiparametric* (next chapter). The term semiparametric refers to the property that the baseline hazard function $h_0(t)$ does not require a parametric definition, but the b_i-coefficients (called *regression coefficients*) are parametric measures that remain in the model to measure the influences of the explanatory variables.

The key feature of a multivariable survival model is that the hazard functions are proportional. Specifically,

$$\text{hazard ratio} = hr_i = \frac{h_i(t \mid x_{i1}, x_{i2}, \ldots, x_{ik})}{h_0(t)} = e^{\sum b_j x_{ij}}.$$

The model dictates that the ratio of hazard functions does not depend on survival time t (proportional = constant hazard ratio). The ratio then exclusively reflects the influences of the explanatory variables and becomes a primary focus of the statistical analysis. Perhaps a more familiar form for the multivariable hazards model is

$$\log \text{hazard ratio} = \log[h_i(t \mid x_{i1}, x_{i2}, \ldots, x_{ik})] = \log[h_0(t)] + \sum b_j x_{ij}.$$

This linear-regression-like expression separates the logarithm of the hazard ratio into an "intercept" term that depends on time and a term that entirely depends on the values of the explanatory variables. A slight modification of this expression,

$$\log \text{hazard ratio} = \log\left[\frac{h_i(t \mid x_{i1}, x_{i2}, \ldots, x_{ik})}{h_0(t)}\right] = \log(hr_i) = \sum b_j x_{ij},$$

shows again that the proportional hazards model requires the influence of the explanatory variables to be determined solely by the model parameters (the b_i-coefficients). Not unlike other regression models, a function of the dependent variable transforms the model into a linear function of the explanatory variables. In the case of a single explanatory variable (x), the log (hazard ratio) becomes a straight line with slope b. In symbols, $h_i(t) = h_0(t)e^{bx_i}$ or $\log[h_i(t)] = \log[h_0(t)] + bx_i$, and clearly the hazard ratio does not depend on the survival time.

Properties

A k-variable proportional hazards model is constructed to have many of the same fundamental properties as most multivariable regression models. Consider two sets of explanatory variables, $x_1, x_2, \ldots, x_k$ and $x'_1, x'_2, \ldots, x'_k$ (for example, two individuals). Two additive proportional hazards models (labeled G and H) based on the two sets of k variables x_{gj} and x'_{hj} are

$$h_G(t \mid x_{g1}, x_{g2}, \ldots, x_{gk}) = h_0(t)e^{\sum b_j x_{gj}}$$

and

$$h_H(t \mid x'_{h1}, x'_{h2}, \ldots, x'_{hk}) = h_0(t)e^{\sum b_j x'_{hj}}.$$

The ratio of these two model hazard functions is

$$\text{ratio of hazard functions} = hr = \frac{h_H(t \mid x'_{h1}, x'_{h2}, \ldots, x'_{hk})}{h_G(t \mid x_{g1}, x_{g2}, \ldots, x_{gk})} = e^{\sum b_j (x'_{hj} - x_{gi})}.$$

Property 1

When all the explanatory variables (x-values) are the same except one (in symbols, $x'_{hj} = x_{gi}$ for all $j \neq m = 1, 2, \ldots, k$), the hazard ratio reflects only the influence of that variable. The influence of the mth explanatory variable is, for example,

$$\text{relative hazard ratio} = hr_m = e^{b_m(x'_{hm} - x_{gm})}.$$

Additive models are designed so that each regression coefficient b_j measures the separate contribution of each variable x_j as if the other $k-1$ explanatory, variables were held constant. Thus, each relative hazard ratio from an additive model indicates the separate influence of each explanatory variable. This property is referred to in a number of ways. It is said that the influence reflected by the coefficient b_j is adjusted for the influences of the other explanatory variables or the influence measured by the coefficient b_j accounts for the influences of the other explanatory variables or the coefficient b_j measures the independent influence of the jth explanatory variable. Such a model is called an additive model because each variable adds its own influence to the survival time, unaffected by the other $k-1$ explanatory variables. For example, when $x'_{hj} - x_{gi} = 1$, the relative hazard ratio ($hr_j = e^{b_j}$) indicates the amount of change in the hazard ratio for a one-unit increase of the jth explanatory variable regardless of the values of the other variables in the model. Additive models are extremely effective statistical tools exactly because of this property.

An ideal comparison occurs when two compared groups are identical in all respects except for one specific variable. Then, logically, any difference observed between the groups is due to that single variable. Assigning individuals at random to one group that receives a treatment and to another that does not receive the treatment (randomization) is an attempt to approximate this ideal comparison. In this case, the groups are not identical but likely balanced with respect to all variables other than the treatment.

The comparison of human survival between nonrandomized groups, however, is frequently far from this ideal. The groups compared typically differ in a number of respects, making it difficult (at best) to attribute observed differences to a single influence (not balanced). An additive model is designed to produce statistical comparisons as if the compared groups were "identical" for all but one variable. Each estimated regression coefficient indicates the influence of a single variable as if the other $k - 1$ variables were balanced between the compared groups, leading to an easily interpreted measure of association. Of course, the imbalances caused by other variables are only "equalized" when they are measured and included in the model. Of equal importance, the model must accurately represent the relationships within the collected data. When randomization is not possible, frequently the situation in the study of human mortality and disease, an additive model provides an opportunity to interpret comparisons between variables as if the groups had been formed by randomization.

Hazard ratios are fundamental and essential summaries of survival data. They are the natural summaries of an additive proportional hazards model. However, hazard ratios are easy to calculate and simple to interpret only when they are estimated from additive models. When, for example, explanatory variables have nonlinear influences or joint influences exist among the explanatory variables (nonadditivity), calculating a hazard ratio becomes more complex and interpretation more difficult.

A remaining issue is the description of an estimated hazard ratio $\hat{hr}_j$ relative to the value one (exactly no influence from an explanatory variable). For example, a relative hazard ratio of 0.2 ($hr_j = h_j(t)/h_0(t) = 0.2$) might be interpreted as an 80% decrease in the hazard rate relative to the "null" value of one. A better description of the decrease is to note that the hazard ratio is five times smaller than the value 1.0. A ratio scale, as the name suggests, measures differences in terms of ratios. For the hazard ratio $hr = 0.2$, the symmetric hazard ratio $1/0.2 = 5$ measures the same degree of influence of an explanatory variable relative to the value 1.0 and is a 400% change, not an 80%. For a multiplicative scale, symmetric means that $1/x$ and x express the same but opposite influences relative to 1.0. That is, the ratios 0.2 and 5.0 both indicate a fivefold change. From another point of view, a property of a ratio scale is that $\log(x)$ and $-\log(1/x)$ reflect the same degree of influence relative to zero but in opposite directions on an additive scale (for example, $\log(5.0) = -\log(0.2) = 1.609$—a 38% change). As always, logarithms convert multiplicative relationships into additive relationships.

Property 2

A multivariable additive hazards model possesses another property common to additive models. The joint influence of two or more additive explanatory variables is a simple function of their separate influences. In the case of a proportional hazards model, the hazard ratio is a product of relative hazard ratios, or

$$\text{hazard ratio} = hr = e^{\sum b_j(x'_{hj}-x_{gj})} = \prod e^{b_j(x'_{hj}-x_{gj})} = \prod hr_j,$$

where, as before,

$$\text{relative hazard ratio} = hr_j = e^{b_j(x'_{hj}-x_{gj})}, \qquad j = 1, 2, \ldots, k.$$

The expression $hr_j = e^{b_j(x'_{hj} - x_{gi})}$ reflects only the influence of the jth explanatory variable in terms of a specific relative hazard ratio and the overall hazard ratio (hr) is the product of k separate relative hazard ratios. When a multivariable summary is a product of a series of individual summaries (one for each measured variable), the influences of the explanatory variables are frequently referred to as *independent*. For example, a measure of the joint influence of the variable g and the variable h on survival time is the product of the relative hazard ratios $hr_g \times hr_h$ when these variables have independent influences. Independence in this context results directly from the model additivity. When the explanatory variables are additive, their influences are independent and vice versa. Independence is the key property of the additive model but is not necessarily a property of the data. Evaluating the degree of correspondence between the relationships postulated by the model (such as independence/additivity) and the relationships within the collected data is an essential part of survival analysis (*goodness-of-fit*) and will be discussed.

Application

The SFMHS includes a cohort of 88 homosexual/bisexual men who were determined to be seropositive (HIV-positive) and entered into a special study (July 1984 to December 1987). The "survival" time for these HIV-positive study participants is the time from entry into the study until the diagnosis of AIDS (in weeks). The end point is illness, not death, but, nevertheless, the length of this AIDS-free time period is referred as a survival time for consistency in terminology. This SFMHS cohort produced 51 complete (AIDS cases) and 37 censored observations (AIDS-free cases) at the end of a 42-month period (3.5 years). The influence of three factors potentially related to the prognosis of AIDS is explored: the $CD4$ lymphocyte count ($CD4$-count), the serum β-microglobulin level (β-level), and the age (age) of the study subject. The units associated with $CD4$-counts are cells/mm^3, but in keeping with the simplest possible terminology, these three covariables are expressed without including the units of measurement. Previous studies of $CD4$-counts and β_2-microglobulin levels described their separate associations with disease severity. Two histograms (Figure 8.1) display the distributions of these two immunologic responses to infection. The following survival analysis also

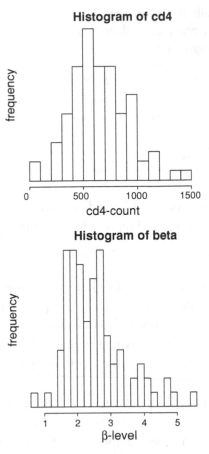

Figure 8.1.   The distributions of *CD4*-counts and β_2-microglobulin levels from the SFMHS HIV/AIDS data.

relates to the significance of these risk measures in the prognosis of the disease, particularly in predicting time-to-AIDS ("survival") based on the extent of infection indicated by the joint influence of the *CD4*-counts and β-levels while accounting for the age of the patient. Throughout the discussion, the multivariable Weibull proportional hazards model is used to identify and assess this joint indication of the level of infection. The exponential hazards model is a special case and its application would follow almost the identical pattern ($\gamma = 1$). A semiparametric analysis of the same data follows in the next chapter.

Table 8.1. Estimates of the Weibull proportional hazards model parameters describing the influence of the reported *CD4*-counts on survival time.

Variables	Symbols	Estimates	Std. errors	p-values
Intercept	b_0	−3.333	—	—
CD4-counts	b_1	−0.002	0.001	<0.001
Shape	γ	1.339	0.747	0.016
		LogLikelihood $= -282.085$		

A single-variable Weibull proportional hazards model representing the relationship among hazard functions and reported *CD4*-counts is

$$h_i(t \mid CD4) = h_0(t) \times c_i = \lambda_i \gamma t^{\gamma-1} = e^{[b_0+b_1 cd4_i]\gamma} \gamma t^{\gamma-1}$$

for the *i*th study participant, where the baseline hazard function is $h_0(t) = e^{b_0 \gamma} \gamma t^{\gamma-1}$, making the hazard ratio $c_i = hr_i = e^{b_1(cd4_i)\gamma}$ per *CD4*-count. As before, the Weibull distribution scale parameter λ_i is constructed to reflect the influence of the explanatory *CD4*-count, where $\gamma_i = e^{[b_0+b_1 cd4_i]\gamma}$. In addition, the shape parameter γ is assumed to be constant (the same for all sampled individuals).

The reported *CD4*-count (denoted $cd4_i$) is entered directly into the hazards model, as reported. This model has the same form as the previously described two-sample proportional hazards model (Chapter 7), but the essentially continuous nature of the *CD4*-count variable gives it a different character. The HIV/AIDS data ($n = 88$ and $d = 51$) yield the maximum likelihood estimated model parameters $\hat{b}_0$, $\hat{b}_1$, and $\hat{\gamma}$ (Table 8.1).

The estimated model coefficient associated with the *CD4*-count ($\hat{b}_1 = -0.002$) translates into the estimated hazard ratio $\hat{hr} = e^{-0.002(1.339)} = 0.997$. The estimated hazard ratio is less than but close to one. A hazard ratio less than one means that lower levels of an explanatory variable produce higher risk (shorter survival times). Thus, the *CD4*-hazard ratio indicates that two persons who differ by a single *CD4*-count have hazard functions that differ by an estimated ratio of 0.997 over the entire range of survival time. Such a small difference might appear inconsequential. However, the model coefficient b_1 measures response per unit change of the explanatory

Table 8.2. Details of a model created to explore the relationships among five nominal categories of CD4-counts (strata).

Interval	z_1	z_2	z_3	z_4	Model
CD4 $\leq$ 450	0	0	0	0	$h(t \mid z_1, z_2, z_3, z_4) = e^{b_0 \gamma} = h_0(t)$
450 $<$ CD4 $\leq$ 600	1	0	0	0	$h(t \mid z_1, z_2, z_3, z_4) = e^{(b_0+b_1)\gamma} = h_0(t)e^{b_1\gamma}$
600 $<$ CD4 $\leq$ 850	0	1	0	0	$h(t \mid z_1, z_2, z_3, z_4) = e^{(b_0+b_2)\gamma} = h_0(t)e^{b_2\gamma}$
850 $<$ CD4 $\leq$ 1000	0	0	1	0	$h(t \mid z_1, z_2, z_3, z_4) = e^{(b_0+b_3)\gamma} = h_0(t)e^{b_3\gamma}$
CD4 $>$ 1000	0	0	0	1	$h(t \mid z_1, z_2, z_3, z_4) = e^{(b_0+b_4)\gamma} = h_0(t)e^{b_4\gamma}$

variable. That is, the magnitude of $\hat{b}_1$ (estimated hazard ratio $= e^{\hat{b}_1 \hat{\gamma}}$) reflects the change in risk per unit (per CD4-count) and $f \times \hat{b}_1$ (estimated hazard ratio $= [e^{\hat{b}_1 \hat{\gamma}}]^f$) reflects the change in risk per f units. For example, two individuals who differ by a CD4-count of $f = 400$ have hazard functions that differ by a ratio of 2.8 [$hr = 0.997^{400} = 0.336 \, (1/0.336 = 2.813)$], a close to three-fold difference in risk.

A fundamental question becomes: Is the postulated linear representation of the influence of CD4-counts (namely, the scale parameter $= \lambda_i = e^{[b_0+b_1 cd4_i]\gamma}$) sufficiently accurate, or will a more sophisticated characterization of the pattern of influence substantially improve the accuracy of the hazards model? An answer to this question potentially emerges from a plot of the coefficients estimated from a model especially constructed to describe a continuous explanatory variable temporarily classified into a sequence of categories (strata).

For the CD4-counts, the sequence of categories chosen is as follows: less than or equal to 450, 450 to 600, 600 to 850, 850 to 1000, and greater than 1000. A variable consisting of k unconstrained categories is referred to as a *nominal variable*. A Weibull hazards model postulated to explore the relationship of these five CD4-categories to survival time based on a nominal variable is

$$h_i(t \mid z_1, z_2, z_3, z_4) = e^{[b_0+b_1 z_{i1}+b_2 z_{i2}+b_3 z_{i3}+b_4 z_{i4}]\gamma} \gamma \, t^{\gamma-1}$$
$$= h_0(t)e^{[b_1 z_{i1}+b_2 z_{i2}+b_3 z_{i3}+b_4 z_{i4}]\gamma}.$$

This hazards model produces five estimated regression coefficients $\hat{b}_i$ (details are in Table 8.2). These coefficients potentially indicate any pattern among

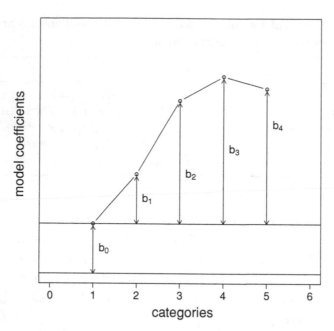

Figure 8.2.   Illustration of a hypothetical pattern of response from a categorical hazards model variable (nominal variable).

the four categories relative to the low $CD4$-count category. The z_{ij}-values are components of a design variable specifically created to allow a k-level nominal variable to be incorporated into the scale parameter of the hazards model. Figure 8.2 illustrates one possible role of each category in the nominal variable model. The first coefficient (b_0) establishes a "baseline" level and the other $k - 1$ coefficients indicate the relative pattern of influence of the explanatory variable. In other words, the amount of change in response associated with each category is measured relative to a referent category. The hypothetical display (Figure 8.2) shows a nonlinear increasing relationship. However, the b_i-coefficients can have any value, in any order.

The estimated influences from the nominal variable (coefficients $\hat{b}_i$) indicate, in a not an extremely efficient fashion (low statistical power), the completely unconstrained relationship between the explanatory variable and survival time in terms of a series of categories. That is, the k coefficients characterize the relationship between a continuous explanatory variable and survival time as well as occasionally suggesting a possible mathematical

Table 8.3. Summary of the application of the Weibull survival model to the five nominal categories of CD4-counts.

Variables	Symbols	Estimates	Std. errors	p-values
Intercept	b_0	3.923	—	—
$450 < CD4 \leq 600$	b_1	−0.573	0.286	0.045
$600 < CD4 \leq 850$	b_2	−0.926	0.287	0.002
$850 < CD4 \leq 1000$	b_3	−0.990	0.398	0.013
$CD4 > 1000$	b_4	−1.111	0.483	0.002
Shape	γ	1.336	0.287	0.313
LogLikelihood = −282.765				

representation. *Unconstrained* means that the relationship can be linear or nonlinear or random or any pattern that can be described with k values. Unlike an additive model, the data entirely determine the observed relationship. Thus, the estimated coefficients associated with the levels of the nominal variable potentially identify a risk/survival time pattern that can be incorporated into a more extensive and efficient hazards model.

The five coefficients estimated from the Weibull proportional hazards model reflect the pattern of influence of the five nominal categories of CD4-counts (Table 8.3) on the time to diagnosis of AIDS. The principal purpose of constructing a nominal variable model is to produce a plot such as Figure 8.3. Visually, the plot (CD4 categories versus $\hat{b}_i$-coefficients) suggests that the originally proposed linear representation of the CD4-counts (dashed line) is not misleading.

When a more complicated relationship is identified, it can be represented by a more sophisticated mathematical expression. For example, the influence of a continuous explanatory variable could be represented by a polynomial expression. Instead of the variable x alone, the three terms x, x^2, and x^3 could be included in the model, potentially improving the model as a representation of the relationship between explanatory variable (x) and survival time. That is, instead of bx, the expression $b_1 x + b_2 x^2 + b_3 x^3$ might more accurately reflect the pattern of influence of the variable x. A linear representation of the influence of the independent variable ($b \times x$) is the simplest possible but not a requirement of an additive model.

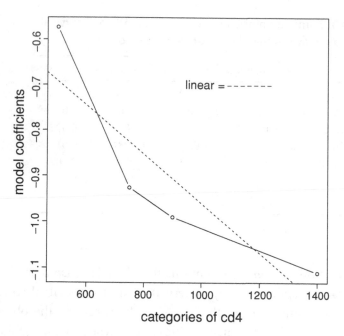

Figure 8.3.   Estimated Weibull hazards model coefficients applied to five nominal categories of *CD4*-counts.

An estimated median survival time based on the additive Weibull hazards model for a specific *CD4*-count is

$$\text{median}(cd4_i) = \hat{t}_m = \left[\frac{\log(2)}{\hat{\lambda}_i}\right]^{1/\hat{\gamma}} = [\log(2)]^{1/\hat{\gamma}} e^{-(\hat{b}_0 + \hat{b}_1 cd4_i)}$$

$$= [\log(2)]^{1/1.339} e^{-(-3.333 - 0.002cd4_i)}.$$

The expression for estimating the median value again is found by solving the expression $\hat{S}(t) = e^{-\hat{\lambda}_i \hat{t}_m^{\hat{\gamma}}} = 0.5$ for the value denoted $\hat{t}_m$. For the example, an individual with the mean *CD4*-count 657.8 ($\overline{cd4} = 657.8$) has an estimated median survival time of $\hat{t}_m = 75.9$ weeks. The estimated median survival time clearly depends on an individual's *CD4*-count. Model estimated median survival times are displayed in Figure 8.4 for a range of *CD4*-counts. The estimated median survival time sharply increases (risk decreases) as the *CD4*-count increases.

The same issue of the appropriate representation of a risk variable arises in constructing a model that includes the influence of β_2-microglobulin

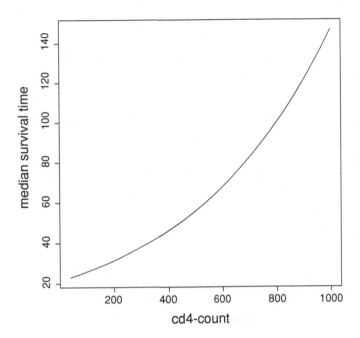

Figure 8.4. Model generated median survival times estimated from the Weibull proportional hazards model (Table 8.1) for a range of *CD4*-counts.

Table 8.4. Estimates of the Weibull proportional hazards model parameters describing the influence of the reported β_2-microglobulin on survival time.

Variables	Symbols	Estimates	Std. errors	p-values
Intercept	b_0	−5.551	—	—
β-levels	b_1	0.371	0.121	0.002
Shape	γ	1.293	0.122	0.035
		LogLikelihood $= -284.656$		

levels. Parallel to the *CD4*-model (linear), directly incorporating the reported β-levels (denoted β_i) into the scale parameter as part of the proportional hazards model

$$h_i(t \mid \beta) = e^{[b_0 + b_1 \beta_i]\gamma} \gamma \, t^{\gamma - 1}$$

produces the maximum likelihood estimated parameters in Table 8.4.

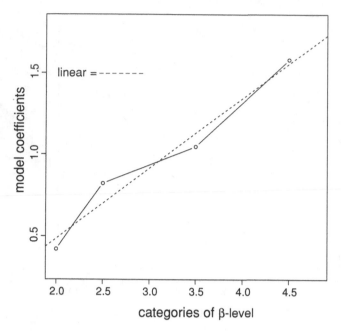

Figure 8.5.   Estimated Weibull hazards model coefficients applied to five nominal categories of β-levels.

Assessing the choice of a linear representation of the β_2-microglobulin influences on survival time follows a pattern similar to assessing the CD4-counts. Again, temporarily categorizing the β-levels (category bounds: 0, 1.75, 2.25, 2.75, 4.25, and 6) produces a plot reflecting its totally unconstrained relationship to survival time (Figure 8.5). A linear representation again appears to summarize accurately the β-level/survival relationship. As with the CD4 plot, no apparent evidence exists to justify a representation of other than the linear influence (scale parameter $= \lambda_i = e^{[b_0 + b_1 \beta_i]\gamma}$).

Again parallel to the CD4 analysis, the median survival times are estimated from the Weibull proportional hazards model (Table 8.4) for a range of β_2-microglobulin levels (Figure 8.6). The median survival time sharply decreases as the β-levels increase.

Multivariable hazards model

An additive multivariable hazards model, as the name suggests, is created from a weighted sum of more than one explanatory variable. The

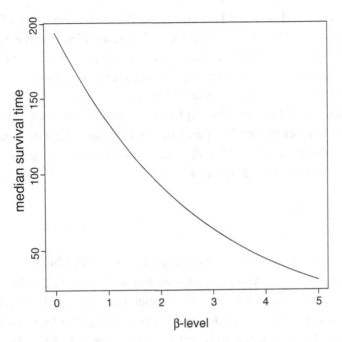

Figure 8.6.   Model generated median survival times estimated from the Weibull proportional proportional hazards model (Table 8.3) for a range of β-levels.

three SFMHS variables $CD4$-counts, β_2-microglobulin, and age from the HIV/AIDS data produce the additive multivariable model, again in terms of proportional hazard functions, given by

$$h_i(t \mid cd4, \beta, age) = h_0(t) \times c_i = h_0(t)e^{b_1 cd4_i + b_2 \beta_i + b_3 age_i}$$

or, for the Weibull hazards model,

$$h_i(t \mid cd4, \beta, age) = e^{[b_0 + b_1 cd4_i + b_2 \beta_i + b_3 age_i]\gamma} \gamma t^{\gamma - 1}$$
$$= h_0(t) e^{[b_1 cd4_i + b_2 \beta_i + b_3 age_i]\gamma},$$

where $h_0(t) = e^{b_0\gamma}\gamma t^{\gamma - 1}$. As before, the scale parameter exclusively reflects the influence of the explanatory variables, namely the scale parameter $=$ $\lambda_i = e^{[b_0 + b_1 cd4_i + b_2 \beta_i + b_3 age_i]\gamma}$. The three explanatory variables are entered into the analysis as reported. In addition, their influences are postulated as additive, producing independent multiplicative changes in the Weibull scale parameter. In the context of a statistical model, additive means (to repeat) that the contribution of each variable to the hazard ratio is reflected by a single parameter b_i and this contribution is unaffected by the levels of the other explanatory variables in the model. For the example, the degree of influence of CD4-counts (b_i) is independent of whether the study subject is young or old. For any age, the CD4-count influence remains entirely determined by the coefficient b_1 and the reported CD4-count.

Models created to describe explanatory variables that do not influence survival in an additive (independent) fashion are called *nonadditive*. An example of a nonadditive Weibull hazards model using the three explanatory measures CD4-counts, β-levels, and age is

$$h_i(t \mid cd4, \beta, age)$$
$$= h_0(t) \, e^{[b_1 cd4_i + b_2 \beta_i + b_3 age_i + b_4 (cd4_i \times \beta_i) + b_5 (cd4_i \times age_i) + b_6 (\beta_i \times age_i)]\gamma}.$$

The explanatory variables are no longer independent. The model allows the relationship between a specific variable and the hazard rate to be influenced by the level of other variables in the model. For example, the three variable nonadditive HIV/AIDS model allows the relationship between the CD4-count and the hazard rate to be influenced by the age of the study subject. The joint influence is more clearly seen from the expression

$$h_i(t \mid cd4, \beta, age)$$
$$= h_0(t) \, e^{[(b_1 + b_5 \times age_i) \times cd4_i + b_2 \beta_i + b_3 age_i + b_4 (cd4_i \times \beta_i) + b_4 (cd4_i \times \beta_i) + b_6 (\beta_i \times age_i)]\gamma},$$

where the relevant model coefficients are b_1 and b_5. The second model is identical to the first nonadditive model, but the variables are rearranged to emphasize the influence of age on the relationship between the CD4-counts and the hazard rate.

Specifically, the influence of the CD4-count on the hazard rate depends on age as long as b_5 is not zero. The influence is determined by the reported CD4-count multiplied by the factor $(b_1 + b_5 \times age)$. For example, if $b_1 = -0.008$ and $b_5 = 0.0002$, then for age 20, the CD4-count is multiplied by 0.003, and for age 40, the same count is multiplied by 0.001. The model dictates that an increase in the study subject's age decreases the influence of the CD4-count on the hazard rate. A twenty-year difference in age reduces the influence of the CD4-count by a factor of 3. Such a nonadditive relationship requires the survival of older individuals to be less influenced by the level of their CD4-count. This kind of joint influence is referred to as nonadditivity or an

Table 8.5. Estimates from the Weibull survival time model applied to the HIV/AIDS CD4-counts, β-levels, and age data including all two-way interactions.

Variables	Symbols	Estimates	Std. errors	p-values
Intercept	b_0	1.923	—	—
cd4	b_1	−0.008	0.0042	—
β	b_2	−0.103	0.829	—
age	b_3	−0.215	0.119	—
cd4 × β	b_4	−0.0004	0.0005	0.525
cd4 × age	b_5	0.0002	0.0001	0.039
β × age	b_6	0.020	0.024	0.423
Shape	γ	1.465	0.120	0.001
		LogLikelihood = −275.705		

interaction or nonindependence or an effect measure modification. However, the model hazard rates remain proportional with respect to time.

The estimated coefficients from the nonadditive hazards model based on the HIV/AIDS data ($n = 88$ and $d = 51$) are given in Table 8.5. The estimated coefficient associated with the CD4/age interaction term ($\hat{b}_5$) indicates that age has a plausibly systematic (nonrandom) influence on the CD4/hazard relationship (Wald's test: $X^2 = (-0.000235/0.000114)^2 = 4.249$, yielding the p-value = 0.039). More generally, the comparison of log-likelihood values produces a similar p-value of 0.042. Formally, the chi-square likelihood ratio test statistic is

$$X^2 = -2[-\log[L_{b_5=0}] - (-\log[L_{b_5\neq0}])]$$
$$= -2[-277.768 - (-275.705)] = 4.127$$

with the resulting p-value $= P(X^2 \geq 4.127 \mid b_5 = 0) = 0.042. = 0.042$. Variables that interact are symmetric in the sense that the CD4-counts can also be viewed as influencing the age/hazard rate relationship. From either point of view, evidence exists that the estimate $\hat{b}_5$ is not likely to be a random deviation from zero.

It is important to keep in mind that the influence from a single explanatory variable is not simply identified in a nonadditive model. Nevertheless, the associated and usually meaningless p-values are frequently given by computer estimation programs. For example, when the interaction term $cd4 \times age$ appears in the model, the single variables $cd4$ and age must also be included, a total of three terms. Thus, two model coefficients are required to measure the influence of an explanatory variable on survival time when a two-way interaction is present. In the presence of an interaction, such as the $cd4 \times age$ interaction, a meaningful summary of the relationship between $CD4$-counts and the hazard rate is not possible with a single coefficient, as indicated by the suppression of the p-values in Table 8.5. For example, the coefficient $\hat{b}_1 = -0.008$ does not directly reflect the influence of the $CD4$-count and the coefficient $\hat{b}_3 = -0.152$ also does not directly reflect the influence of age (Table 8.5). An interaction means that an unbiased estimate of the influence of the $CD4$-count is accurately achieved only for a specific age. A single coefficient summarizes the role of a single variable only when it has an additive influence. In practical terms, a test of the absence/presence of an interaction addresses the question: Is a single regression coefficient sufficient to summarize accurately the influence of a specific variable on survival time?

When a statistical model contains nonadditive terms, adjusted coefficients are not an issue. Nonadditivity means that a summary based on a single variable is misleading. Therefore, the degree to which the influence on survival time from a single variable is affected by another explanatory variable or variables (confounding) is moot. In other words, when an interaction exists, confounding is not a relevant issue. Conversely, an additive model provides the opportunity to explore separately and parsimoniously the influence of each explanatory variable. Adjustment for the influence of other explanatory variables then becomes a relevant and important issue. This fundamental property of a multivariable model (independence?) makes the choice between describing the survival data with a nonadditive or an additive relationship critically important to the analysis.

A word of warning is worthwhile at this point. Two kinds of errors can be made. Nonadditive terms can be included in the model when they are not necessary or nonadditive terms can be omitted from the model when they are necessary. The former error is not serious. The later error is potentially devastating. "Wrong model bias" incurred by not including necessary nonadditivity produces at best approximate and at worst entirely misleading results (many examples exist).

A first step in describing the pattern of nonadditivity identified with a multivariable model is to stratify the data by one of the two nonindependent

Table 8.6. The relationships between CD4-count and the hazard rate (model coefficients) and their approximate 95% confidence interval bounds for four age categories (strata).

Age categories	n	Estimates	Lower	Upper
age $\leq$ 25	18	−0.0036	−0.0064	−0.0008
25 < age $\leq$ 30	22	−0.0025	−0.0048	−0.0001
30 < age $\leq$ 35	32	−0.0020	−0.0034	−0.0006
age > 35	16	−0.0010	−0.0002	0.0014

variables. For the HIV/AIDS example, four strata are created by temporarily categorizing the age variable to explore the CD4/age interaction. Parallel to the previous nominal variable models, the creation of temporary sequence of age categories ($\leq$ 25, 25 to 30, 30 to 35, and $\geq$ 35) and a stratified analysis potentially reveals the pattern of nonadditivity. The model coefficients reflecting the influence of the CD4-count estimated within each of the four age groups are given in Table 8.6. Again, the analysis employing categorical variables is not a powerful approach but possibly suggests model-free patterns of influence and, for the HIV/AIDS-data, the possible pattern of the CD4/age interaction.

From the analysis based on stratified age categories (Table 8.6), the CD4-count influence on the hazard rate appears similar (little or no interaction) for individuals less than 35-years old ($age \leq 35$). For these men, their CD4-count have a more or less constant influence on the hazard rate ($\hat{b}$-coefficients in the neighborhood of −0.0025). The influence of the CD4-count observed in men older than 35 years ($\hat{b}$-coefficient $= -0.0010$) is considerably smaller. Essentially the same result is achieved by applying the Weibull proportional hazards model made up of the CD4-counts (as reported) with age as a nominal variable and including all two-way interactions.

The four analyses summarized in Table 8.6, despite the unavoidably small numbers of observations, indicate that an additive hazards model likely produces an accurate and potentially useful summary of the relationship between the CD4-count and survival time for men under 36 years of age. Such an additive Weibull proportional hazards model is (repeated)

$$h_i(t \mid cd4, \beta, age) = e^{[b_0 + b_1 cd4_i + b_2 \beta_i + b_3 age_i]\gamma} \gamma t^{\gamma - 1}.$$

Table 8.7. Estimates from the additive Weibull proportional hazards model based on the variables CD4-counts, β-levels and age ($n = 72$, age ≤ 35).

Variables	Symbols	Estimates	Std. errors	p-values
Intercept	b_0	-3.439	—	—
cd4	b_1	-0.002	0.001	0.005
β	b_2	0.265	0.123	0.031
age	b_3	-0.010	0.028	0.726
shape	γ	1.309	0.130	0.038
		LogLikelihood $= -232.281$		

The estimated coefficients (Table 8.7) define this additive model for the HIV/AIDS data, excluding the 16 study subjects who are older than 35 years ($n = 72$).

For an additive model, the first analytic issue is the assessment of the impact of random variation on the estimated model coefficients. Test statistics (Wald's tests, for example) and their p-values indicate the variables that likely have nonrandom influences on survival time. Both the CD4-counts (p-value $= 0.005$) and the β-levels (p-value $= 0.031$) likely influence the hazard rate in a systematic way. No similar evidence exists for the influence of age (p-value $= 0.729$). In addition, the estimated Weibull shape parameter ($\hat{\gamma} = 1.309$) indicates that a description of survival time based on the simpler exponential hazards model ($\gamma = 1$) would be inadequate (p-value $= 0.039$).

An additive model, because it requires each explanatory variable to have a separate influence on survival time, is frequently an excellent statistical tool for clearly describing the relationships within the sampled data. The model estimates simply describe the relative and independent contributions to the overall hazard rate (risk) from each explanatory variable. Direct comparisons among the coefficients, however, fail to produce profitable information when the variables are measured in different units. For example, the CD4-count is not less important than the age in the prediction of a hazard rate because the estimated regression coefficient $\hat{b}_1 = -0.002$ (cd4) is closer to zero than the coefficient $\hat{b}_3 = -0.010$ (age). Measurement units largely determine the magnitude of a regression coefficient. When a study participant's age is measured in months rather than years, for example, the coefficient $\hat{b}_3$ decreases by a factor of 12 (new-$\hat{b}_3 = -0.001$).

Dividing estimated coefficients by their standard errors creates unitless values providing direct and sometimes useful comparisons. The standardized coefficients are said to be *commensurate* (same units). Again, these comparisons require the coefficients to be estimated from an additive model.

For the AIDS data, commensurate statistical summaries are as follows:

$$c d4\text{-counts: } \hat{B}_{cd4} = \frac{\hat{b}_1}{S_{\hat{b}_1}} = \frac{-0.00228}{0.00065} = -3.479,$$

$$\beta\text{-levels: } \hat{B}_\beta = \frac{\hat{b}_2}{S_{\hat{b}_2}} = \frac{0.265}{0.123} = 2.157$$

and

$$\text{age: } \hat{B}_{age} = \frac{\hat{b}_3}{S_{\hat{b}_3}} = \frac{-0.010}{0.028} = 0.350.$$

For these men (age less than 35 years), the level of $CD4$-counts has the most influence on the estimated hazard ratio, followed by an important but smaller influence from the level of β_2-microglobulin. Their age has a relatively small and likely random influence. The Wald test-statistics and the p-values associated with these summaries are also commensurate measures and comparisons reflect the relative influences. In fact, a number of ways exist to make these kinds of commensurate comparisons among explanatory variables and typically produce similar but not identical results.

The defining feature of an additive model is a simple description of the joint influences of the explanatory variables. For example, the joint influences of $CD4$-counts, β-levels, and age are effectively described by the median survival time (time scale) based on the estimated Weibull proportional hazards model, where

$$\text{median value} = \hat{t}_m = [\log(2)]^{1/\hat{\gamma}} e^{-(b_0 + \sum b_j x_{ij})}$$

$$= [\log(2)]^{1/\hat{\gamma}} e^{-(b_0 + b_1 cd4_i + b_2 \beta_i + b_2 age_i)}$$

$$= [\log(2)]^{1/1.309} e^{-(-3.439 + 0.002 cd4_i + 0.265\beta_i - 0.010 age_i)}.$$

For HIV-positive men of age 28.5 years (mean value), estimated median values describe the joint influences on survival time from nine selected pairs of $CD4$-counts and β-levels (Table 8.8). That is, these model-generated summary values provide a rigorous and intuitive description of the degree of joint response to infection (time scale—weeks).

Table 8.8. The joint influence (time scale) of CD4-counts and β_2-microglobulin in terms of estimated median time to AIDS (age = 27.5 years).

	β_2-microglobulin levels		
	1.5	2.5	3.5
cd4-counts = 400	52.1	39.9	30.7
cd4-counts = 700	103.0	79.1	60.7
cd4-counts = 1000	203.9	156.5	120.0

Table 8.9. The joint influence (ratio scale) of CD4-counts and β_2-microglobulin in terms of estimated hazard ratios (age = 27.5 years).

	β_2-microglobulin levels		
	1.5	2.5	3.5
cd4-counts = 400	1.73	2.44	3.46
cd4-counts = 700	0.71	1.00	1.41
cd4-counts = 1000	0.29	0.41	0.58

Note: Hazard ratios are relative to CD4-count = 700, β-level = 2.5 for age = 28.5.

Another natural description of the joint influence of the three explanatory variables is the hazard ratio (ratio scale). As will be described (Chapter 9), hazard ratios can be estimated without specifying a specific parametric model (baseline hazard function). However, based on the Weibull additive model (age $\leq$ 35), the overall hazard ratio (product of variable specific relative hazard ratios) directly calculated from the estimated additive model regression coefficients for the ith individual is

$$\hat{hr}_i = \frac{\hat{h}_i(t)}{\hat{h}_0(t)} = e^{b_i(cd4_i)\hat{\gamma}} \times e^{b_2(\beta_i)\hat{\gamma}} \times e^{b_3(age_i)\hat{\gamma}}$$

$$= e^{-0.002(cd4_i)1.309} \times e^{0.265(\beta_i)1.309} \times e^{-0.010(age_i)1.309}$$

$$= 0.997^{cd4_i} \times 1.417^{\beta_i} \times 0.987^{age}.$$

Again for HIV-positive men aged 28.5 years, Table 8.9 contains the estimated joint influences of the previous nine selected pairs of values of CD4-counts and β-levels described in terms of hazard ratios relative to an individual with a $CD4$-count of 700 and a β-level of 2.5 ($hr = 1.0$). A little algebra shows that the estimated median and hazard ratio are related.

Goodness-of-fit

Once a model hazard function is estimated, the corresponding estimated survival function directly follows. In general, the relationship is, as follows:

when $h_i(t) = h_0(t) e^{\sum b_j x_{ij}}$,   then $S_i(t) = [S_0(t)]^{\exp(\sum b_j x_{ij})}$.

Therefore, an estimate of the baseline survival function $\hat{S}_0(t)$ and a set of explanatory variables produce an estimate of the survival function for a specific person or group.

To illustrate, consider the exponential proportional hazards model. The general hazards model is

$$h_i(t) = h_0(t) e^{\sum b_j x_{ij}}.$$

Because the hazard rates ($h_i(t) = \lambda_i$) are constant (exponential survival),

$$\lambda_i = \lambda_0 e^{\sum b_j x_{ij}},$$

multiplying by the survival time t and exponentiating gives

$$e^{-\lambda_i t} = \left[e^{-\lambda_0 t} \right]^{\exp(\sum b_j x_{ij})} \quad \text{and} \quad S_i(t) = [S_0(t)]^{\exp(\sum b_j x_{ij})}.$$

For the Weibull proportional hazards model, similarly the survival function is

$$S_i(t) = \left[e^{-\lambda_0 t^\gamma} \right]^{\exp([\sum b_j x_{ij}] \gamma)}, \qquad \text{where } \lambda_0 = e^{b_0 \gamma}.$$

The baseline scale parameter λ_0 and the shape parameter γ are estimated from the data.

For the SFMHS HIV/AIDS data, each individual observation (survival time t plus specific values of the $CD4$-, β-, and age-variables) generates an estimated Weibull survival probability given by

$$\hat{S}_i(t) = \left[e^{-0.011 t^{1.309}} \right]^{\exp([-0.002 cd4_i + 0.265 \beta_i - 0.010 age_i] 1.309)}$$

and $S_0(t) = e^{\hat{\lambda}_0 t^\gamma} = e^{-0.011 t^{1.309}}$, where $\hat{\lambda}_0 = e^{-3.439(1.309)} = -0.011$.

The expression for $\hat{S}_i(t)$ shows that proportional hazard functions dictate that the survival functions do not intersect. Thus, a plot of the product-limit estimated survival functions provides a preliminary visual evaluation of proportionality but certainly not a conclusive one. Random variation can cause survival functions to cross when the underlying hazard functions are proportional and not to cross when the underlying hazard functions are not proportional.

For all regression models, a specifically designed statistic to evaluate the accuracy of a postulated model (goodness-of-fit) is called a *residual value*. Residual values reflect the difference between the model-estimated values and the observed data that generated the model estimates. Typically a residual value of zero occurs when a model generated value perfectly corresponds to an observed value. Otherwise, residual values indicate what is not reflected by the model. When a model "fits" the data, the residual values are likely small and certainly randomly distributed. When a model fails to "fit" the data in some systematic way, at least some of the residual values are likely large and nonrandom. Occasionally, their observed pattern indicates the specific reason for the "lack of fit."

Statistical justification of the properties of residual values is occasionally mathematically complex and the more complex details are not presented. Assessment of the model's fit based on residual values, however, is not complicated and is simply applied without detailed knowledge of their mathematical origins. The final product is frequently a plot designed to produce an easily interpreted visual picture of the correspondence between the statistical model and the observed data (introduced in Chapters 5 and 6). Using the HIV/AIDS data ($CD4$-counts, β-levels, and age) and the estimated Weibull hazards model, a description follows of two goodness-of-fit techniques.

A simple transformation of the survival function $S(t)$ yields residual values with several special and advantageous properties. The transformation is $r_i = -log[\hat{S}(t_i)]$, called the *Cox–Snell residual value*. In more detail, the Cox–Snell residual value r_i is

$$r_i = -\log[\hat{S}_0(t_i)] = -\log[\hat{S}_0(t_i)]e^{[\sum b_j x_{ij}]}.$$

These r_i-values apply to survival distribution models in general. Therefore, for a specific survival model, the estimated baseline survival function $\hat{S}_0(t)$ and the estimated coefficients $\hat{b}_j$ allow a residual value r_i to be calculated for each survival time t_i and each set of explanatory variables $\{x_{i1}, x_{i2}, \ldots, x_{ik}\}$.

For the Weibull proportional hazards model, the Cox–Snell residual values are

$$r_i = [\hat{\lambda}_0 t_i^{\hat{\gamma}}] e^{[\sum b_j x_{ij}]\hat{\gamma}}$$

using the hazards model estimates λ_0, γ and the k estimated regression coefficients $\hat{b}_j$.

The more familiar residual values resulting from the application of a linear regression model always have a mean value of exactly zero and any variance; they take on any value (in mathematics, it is said that they range from minus infinity to plus infinity) and are uncorrelated with the predicted value. Perfect correspondence between an estimate and an observed values produces a residual value of exactly zero. When a linear model is appropriate, plots of these residual values lead to symmetric displays. The Cox–Snell residual values have none of these properties. Their mean value and variance depend on the number of randomly censored observations, they take on only positive values, and they typically have an asymmetric distribution. These residual values are never zero. When the survival model is appropriate, plots of these residual values do not lead directly to symmetric displays. The role of these residual values in evaluating an estimated model, however, is not different in principle from that of the more typical residual values created to evaluate most regression models.

Residual values estimated from all statistical models have a property in common. When the model is satisfactory, the properties of the residual values are known. An important property of the Cox–Snell residual values is that they are a random sample from an exponential distribution (Chapter 5) with parameter $\lambda = 1$ when the estimated survival model $\hat{S}(t)$ accurately reflects the postulated survival model $S(t)$. In symbols, the probability distribution of these residual values r becomes $G(r) = P(R \geq r) = e^{-r}$ when the survival functions $\hat{S}(t)$ and $S(t)$ differ by chance alone.

For survival models such as the Weibull proportional hazards model, a residual value can be calculated for all observations (complete and censored). A residual value for each observation is defined as

$m_i = r_i$ when the observation t_i is complete

and

$m_i = r_i + 1$ when the observation t_i is censored.

Because the r_i-values have an exponential distribution, the residual values associated with censored values are made "complete" on average by adding the mean value $\mu = 1/\lambda = 1.0$. The same strategy was used (Chapter 5) to estimate the mean value of exponentially distributed survival times. When a variable has an exponential distribution, all randomly censored observations have the same mean value as the observed values, namely μ or, in the case of residual values, $\mu = 1.0$. Therefore, a single residual m_i-value exists for each observation. These complete and "completed" Cox–Snell residual values are called *modified Cox–Snell residual values*.

When the estimated hazards model is appropriate, the modified Cox–Snell residual values are random with respect to the variables used in the analysis ("no more remains to be explained"). When the model fails to capture aspects of an explanatory variable or variables, the pattern associated with the modified Cox–Snell residual values will likely be noticeably nonrandom. Plots of the modified Cox–Snell residual values, therefore, potentially reveal evidence of lack of fit. As mentioned, observed patterns may also suggest ways to improve the correspondence between model and data.

Application

Three plots of the modified Cox–Snell residual values (m_i) from the three-variable additive Weibull hazards model (Table 8.7) are displayed in Figure 8.7. The m_i values plotted against the CD4-counts appear to have a nonrandom pattern but with no apparent systematic features. The β-level and age plots show no obvious indication of nonrandomness among the residual values. In general, no reason exists that the variables plotted have to be those included in the hazards model. The appearance of a pattern of modified residual values associated with a variable not in the model indicates a possible role in the study of survival times.

Another assessment of model "fit" based on the Cox–Snell residual values employs a transformation that produces a straight line. A straight line provides a clear and an intuitive comparison to a straight line theoretically expected when the model exactly reproduces the data. Rupert Miller [5] states in his text on survival analysis,

Basic Principle. Select the scales of the coordinate axes so that if the model holds, a plot of the data resembles a straight line, and if the model fails, a plot resembles a curve.

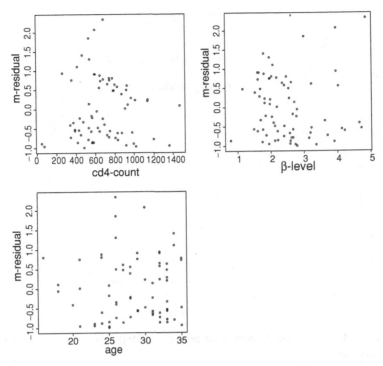

Figure 8.7.   Modified Cox–Snell residual plots for the explanatory variables CD4-counts, β-levels, and age from the Weibull proportional hazards additive model (Table 8.7).

When the data are accurately represented by the estimated hazards model, as stated without justification, the residual values are a random sample from a "unit" exponential probability distribution (denoted previously as $G(r) = P(R \geq r) = e^{-r}$). As noted, log–log transformed exponentially distributed survival probabilities randomly deviate from a straight line with intercept $\log(\lambda)$ and slope of 1 (Chapter 5). Thus, log–log transformed residual values plotted against the logarithms of the residual values themselves will randomly deviate from a straight line (intercept $= \log(\lambda) = \log(1) = 0$ and slope $= 1$) when the estimated model "fits" the data. By "fits," as before, is meant that the model generated values and the corresponding observed values differ only because of random variation. It has been suggested that the accuracy of this kind of assessment of the residual values is not reliable for small samples of data making it most effective when the sample size is large [2].

The plots of the estimated probabilities (denoted $\hat{G}[r_i]$) associated with each residual value (r_i) are displayed in Figures 8.8 and 8.9 for the SFMHS

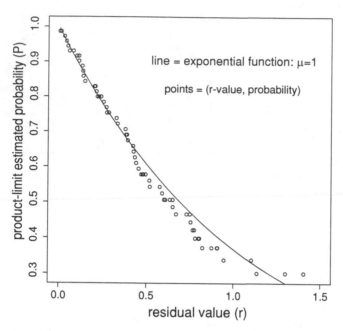

Figure 8.8. Plot of the Cox–Snell estimated probabilities ($\hat{G}[r_i]$) against the residual values (r_i)– exponential?

HIV/AIDS data. Figure 8.8 is a direct plot of the estimated probabilities ($\hat{G}[r_i]$) and the Cox–Snell residual values (exponential?). Figure 8.9 is the plot of the "log-log" transformed estimated probabilities and the logarithm of the Cox–Snell residual values (a straight line with intercept $= 0$ and slope $= 1$?). The plotted transformed residual values in Figure 8.9 appear to randomly deviate from a straight line (estimated intercept $= 0.003$ and slope $= 0.962$), producing no apparent evidence that the additive Weibull hazards model based on three explanatory variables (Table 8.7) can be substantially improved. That is, the exponential distribution ($\lambda = \mu = 1$) appears to be an accurate description of the residual values, which occurs when the estimated hazards model is a useful description of the sampled survival data.

Nonrandom residual values might indicate that additional explanatory variables are necessary or that more sophisticated representations of specific explanatory variables are required or, of course, both. For example, instead of an explanatory variable represented by x (linear), explanatory variables such

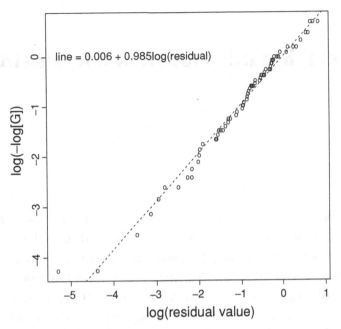

Figure 8.9.   Plot of the transformed Cox–Snell estimated probabilities ($\log(\hat{G}(r_i)])$) against the logarithm of the residual values ($\log[r_i]$)–linear?

as $ax + bx^2 + cx^3$ or $\sqrt{x}$ or $\log(x)$ or $x\log(x)$ might better characterize the relationship of an explanatory variable to survival time improving a model's "fit." Also, for additive models, a potential source of lack of fit is the failure of the explanatory variables to be additive (failure to include interactions terms in the model).

A substantial weakness exists in evaluating survival models using residual values and their plots, such as Figures 8.7, 8.8, and 8.9. There is no objective means to unequivocally identify deficiencies visually. In addition, even when nonrandom elements are identified, no guarantee exists that the model can be substantially improved. The truth of the matter is that few rules exist to guide the creation of an appropriate survival time model. Subject matter considerations plus observing changes in residual value patterns from a trial and error approach are typically at the center of developing successful statistical models. Although not emphasized in most statistical texts, model selection is largely a subjective process. Frank Harrell [6] notes, "Using the data to guide the data analysis is almost as dangerous as not doing so."

General hazards model: nonparametric

Introduction

When the underlying statistical structure of a sampled population is not accurately described by a postulated probability distribution or is simply unknown, a parametric hazards model is not a feasible analytic approach to describing survival time data. Statistician David Cox has provided an alternative. Instead of a fully defined parametric model (Chapter 8), he developed a method of estimating the influences of the explanatory variables that does not require a baseline hazard function to be specified. Furthermore, the baseline hazard function is not much of an issue in many survival analyses, where the main focus is on the role of the explanatory variables, making this distribution-free approach an important statistical tool for analyzing survival data.

The form of the multivariable hazards model remains the same as in the parametric case, where

$$h_i(t) = h_0(t) \times c_i = h_0(t)e^{\sum b_j x_{ij}}.$$

The model hazard functions are again proportional. However, the model regression coefficients (represented by b_j) are estimated without assumptions about the baseline hazard function $h_0(t)$ or the population that generated the data. The Cox approach is said to be "robust" because it applies to a wide variety of situations.

As noted, the estimation process is not assumption-free. It remains required that the sampled hazard functions be proportional. As before, a key component of the analysis is the answer to the question: Are the relationships within the sampled data accurately represent by proportional hazard functions? The application of the Cox semiparametric technique, like most model-based techniques, requires an assessment of the goodness-of-fit. In addition, censoring of survival times is still required to be noninformative.

An important role of nonparametric methods is the comparison to the parametric methods applied to the same data. Previously, the survival curve estimated from a nonparametric approach (product-limit estimate) was compared to the survival curve estimated based on the parametric assumption that the data were sampled from an exponential distribution (Figure 5.4). The degree of success of the parametric model was apparent from a plot of these two estimates. A similar comparison was made for the Weibull estimated survival curve (Figure 6.6). The comparison of the Cox estimated hazards model to the parallel parametric approach is similarly useful as part of a strategy for choosing an appropriate hazards model. If the semiparametric and fully parametric analyses produce close to the same results, the richer, simpler, and more efficient parametric model typically becomes the better choice. Conversely, when the two approaches differ substantially, the parametric model has failed in some important respect. The strategy then becomes to fix the parametric model or to use the Cox model instead.

Obviously, not all survival times are accurately represented by proportional hazard functions. Two commonly encountered situations that are not likely to produce proportional hazard functions are displayed in Figures 9.1 and 9.2.

Figure 9.1 displays two hazard functions that might describe survival data from critically ill patients, such as those who need an organ transplant, where two choices can arise: a high-risk surgical procedure or the usual treatment. The hazard function reflecting the transplant patients' survival would increase initially because of the high risk of the surgery, followed by relatively constant risk. The "usual-care" patients' survival would likely be described by a steadily increasing hazard function over time (continually increasing risk). The ratio of these two kinds of hazard functions would not be constant (Figure 9.1).

Figure 9.2 displays two hazard functions that might describe survival data from a clinical trial. To evaluate a new treatment, a clinical trial starts with patients randomly divided into two groups. One group receive a treatment and the other serve as controls. At the beginning of the trial, both groups

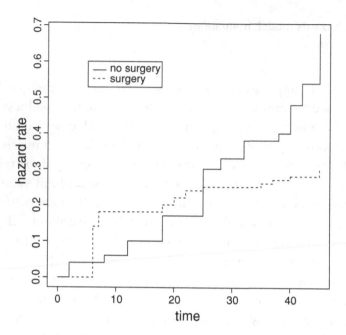

Figure 9.1. Hypothetical example of two nonproportional hazard functions—different patterns of increasing risk.

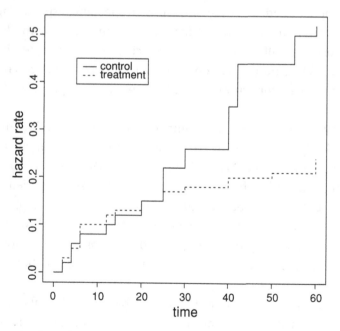

Figure 9.2. Hypothetical example of two nonproportional hazard functions—the same risk at the beginning of a randomized trial (time = 0).

have similar risks because of the original randomization. However, if the new treatment increases survival time, the group receiving the treatment will experience a hazard rate that increases at a slower rate. Again, a direct application of a proportional hazards model would fail to accurately reflect the risk/survival relationship (Figure 9.2).

Cox estimation for proportional hazards model: two-sample case

The two-sample proportional hazards model provides a simple introduction to the estimation and interpretation of the semiparametric approach. The two-sample proportional hazards model is again

$$h_F(t) = h_0(t) \times c = h_0(t) \times e^{bF},$$

where, as before, the symbol F represents a binary variable (coded 0 or 1), indicating two sources of survival data. Using a process called *partial likelihood estimation*, the single coefficient b and its variance are estimated based on this proportional hazards model and, to repeat, no assumption is made about the parametric form of the baseline hazard function $h_0(t)$.

Using the AIDS survival data classified into nonsmokers and smokers ($n = 23$ African-Americans—Chapter 7) and the assumption that $h_0(t)$ is a Weibull baseline hazard function, an estimate of the hazard ratio is $\hat{hr} = e^b = e^{-0.759} = 0.468$ ($S_{\hat{b}} = 0.557$). The same parameter, estimated without specifying the baseline hazard function (partial likelihood estimate), is $\hat{hr} = e^b = e^{-0.823} = 0.439$ ($S_{\hat{b}} = 0.592$). Like maximum likelihood estimates, the estimated standard error is part of the partial likelihood estimation process.

Although partial likelihood estimates are nonparametric, they have the same kinds of properties as maximum likelihood estimates. Therefore, the interpretation and evaluation of the estimated model components are essentially the same as for other regression models. Thus, the partial likelihood estimate $\hat{b}$ has an approximate normal distribution as long as the sample size is moderately large, making it possible to assess the influence of sampling variation directly with the usual statistical tools (tests and confidence intervals). For example, the previous Wald test statistic from the African-American smoking data ($b = 0$?),

$$X^2_{Wald} = \left[\frac{\hat{b} - 0}{S_{\hat{b}}}\right]^2 = \left[\frac{-0.823 - 0}{0.592}\right]^2 = (-1.390)^2 = 1.933,$$

has an approximate chi-square distribution with one degree of freedom when $h_1(t) = h_0(t)$ or $b = 0$. The associated p-value $P(X^2 \geq 1.933 \mid b = 0) = 0.164$ yields no persuasive evidence of an important difference in survival times between nonsmokers and smokers. The similar result from the parametric-based Weibull distribution analysis is

$$X^2 = \left[\frac{\hat{b} - 0}{S_{\hat{b}}}\right]^2 = \left[\frac{-0.759 - 0}{0.557}\right]^2 = (-1.362)^2 = 1.854$$

and the p-value is 0.173.

Again, parallel to the maximum likelihood approach, the difference in partial log-likelihood values calculated from two nested models (multiplied by -2) produces a test statistic with an approximate chi-square distribution when only random differences exist between the compared samples of survival times. For the comparison of the African-American nonsmokers and smokers, the difference in log-likelihood values,

$$X^2_{likelihood} = -2[\log(L_{b=0}) - \log(L_{b\neq0})] = -2[-38.422 - (-37.517)] = 1.809,$$

has an approximate chi-square distribution with one degree of freedom. The probability, that this difference arose by chance alone is p-value $= 0.179$. As previously noted, these two approaches (Wald and likelihood statistics) generally produce similar results, particularly when large samples of data are available from sampled populations. However, it should be noted that when a Wald test statistic is extremely large (evidence that the null hypothesis is extremely likely to be false), its distribution is no longer accurately approximated with a chi-square distribution. Because this inaccuracy is not a property of the likelihood approach, it is generally the preferred statistical assessment.

A log-rank test (Chapter 7) addresses the same question: Do the survival times of nonsmokers and smokers systematically differ? This assessment of a two-sample difference also does not require assumptions about the exact nature of the underlying survival distribution, particularly the hazard functions. The statistical summaries from log-rank test applied to the African-American AIDS data are as follows (repeated from Chapter 7):

1. $\sum a_i = 5$—the total deaths observed among smokers,
2. $\sum \hat{A}_i = 2.898$—the total deaths among smokers estimated as if smoking exposure were entirely unrelated to survival time, and
3. $\sum v_i = 2.166$—the variance of the summary statistic $\sum a_i$.

The chi-square distributed test statistic is

$$X^2_{\text{log-rank}} = \frac{\left(\sum a_i - \sum \hat{A}_i\right)^2}{\text{variance}\left(\sum a_i\right)} = \frac{(5 - 2.898)^2}{2.166} = 2.040.$$

The three approaches to comparing survival times between nonsmokers and smokers among the 23 African-American study participants give similar results ($X^2_{\text{Wald}} = 1.93$, $X^2_{\text{likelihood}} = 1.81$, and $X^2_{\text{log-rank}} = 2.04$).

The log-rank test is a special case of comparing partial likelihood values calculated from a two-sample proportional hazards model. Thus, these two seemingly different techniques always give similar results. The log-rank test and a slightly different likelihood procedure, called a *score likelihood test*, are algebraically identical when all sampled survival times differ ("no ties"). A rigorous demonstration of the equality of these two procedures exists in several more theoretical survival analysis texts [5].

The fact that the log-rank test is a special case of the two-sample Cox approach gives a hint of the underlying process that produces the estimate $\hat{b}$ without specifying the parametric form of the hazard function. The log-rank test begins with a measure of the association between a risk factor and an outcome estimated within each of a sequence of strata, namely $a_i - A_i$. Each stratum estimate is based only on survival times of equal lengths. The log-rank test statistic X^2 combines these stratum-specific measures to create an overall single summary of the risk/outcome association that is, therefore, not influenced by survival time or noninformative censored observations. No assumptions are made about the sampled populations (nonparametric).

A similar process creates the estimate of the coefficient b using the Cox partial likelihood approach. For each stratum, again based on the same survival times, an estimate of the parameter b is made. These stratum-specific estimates of the association between a risk factor and an outcome are also combined to create an overall single summary, namely $\hat{b}$. And, like the log-rank procedure, the overall estimate is not influenced by survival time and is not biased by noninformative censored observations.

The estimate $\hat{b}$ reflects the same hazard ratio at all survival times as long as the compared hazard functions are proportional. Thus, for the smoking data, using the estimate $\hat{b} = -0.823$, the

$$\text{hazard ratio} = \hat{c} = e^{\hat{b}} = e^{-0.823} = 0.439 \quad \text{or} \quad 1/0.439 = 2.278$$

estimates this single constant ratio.

A general property of the Cox estimation process is evident from the two-sample smoking data. The hazard ratio is readily estimated but, in contrast to the Weibull and exponential models, estimates of the specific components of the ratio are not available. For example, under the assumption of a sampled Weibull distribution, the estimated scale parameters from the smoking data are as follows: for nonsmokers, $\hat{\lambda}_0 = 0.00678$, and for smokers, $\hat{\lambda}_1 = 0.0145$, with estimated shape parameter $\hat{\gamma} = 1.421$. Furthermore, the estimated hazard ratio $\hat{hr} = e^{0.759} = 2.136$ is $\hat{hr} = 0.0145/0.00678 = 2.136$. The lack of similar estimates of the specific components of the hazard ratio is a cost incurred by the Cox partial likelihood approach.

As with the previous maximum likelihood estimated model coefficients, the Cox partial likelihood estimated coefficients and variances also produce approximate 95% confidence intervals. From the smoking data, such a confidence interval is

$$\hat{b} \pm 1.960 S_{\hat{b}} = -0.823 \pm 1.960(0.592),$$

which produces the 95% bounds $(-1.984, 0.337)$ for the parameter b. The approximate 95% confidence interval becomes $(e^{-1.984}, e^{0.337}) = (0.137, 1.401)$ for the underlying hazard ratio hr estimated by $\hat{hr} = e^{-0.823} = 0.439$. This confidence interval is consistent with the previous assessments of the influence of random variation on the estimated coefficient $\hat{b}$. The parameter value $b = 0$ ($hr = 1$) is contained in the 95% confidence interval.

A two-sample model is readily extended. The addition of a second explanatory variable (denoted x) creates the proportional hazards model

$$h_i(t \mid F, x) = h_0(t)e^{b_1 F + b_2(x_i - \bar{x})}.$$

Table 9.1. Estimated coefficients from the SFMHS data ($n = 72$ and age ≤ 35) for three models (estimated standard errors in parentheses).

Coefficients	Interaction model	Additive model	Two-sample model
b_1	−0.904 (0.335)	−0.898 (0.335)	−0.873 (0.334)
b_2	−0.052 (0.044)	0.026 (0.035)	—
b_3	0.072 (0.075)	—	—
LogLikelihood	−158.827	−159.301	−159.559

Again, using the HIV-positive subjects from the SFMHS data ($n = 72$, age ≤ 35) to illustrate, a binary $CD4$-variable is defined as "normal" when $CD4 \geq 700$ ($F = 0$) and "low" when $CD4 < 700$ ($F = 1$). In addition, the study subject's age is included in the model as reported ($x_i = age_i$). The specific proportional hazards model becomes

$$h_i(t \mid F, age) = h_0(t)e^{b_1 F + b_2(age_i - \overline{age})}.$$

The "centered" variable ($age_i - \overline{age}$) produces a useful interpretation of the baseline hazard function. The baseline hazard function $h_0(t)$ reflects risk for "normal" $CD4$-level individuals ($F = 0$) who are of average age ($age = 28.583$). In symbols, $h_i(t \mid F = 0, age_i = 28.583) = h_0(t)$. Occasionally computational advantages exist to centering the variables in a proportional hazards model. Using centered variables in an additive model changes the estimated regression coefficients but does not affect the statistical analysis (tests or p-values or inferences).

Three hazards models important in exploring the relationship of $CD4$-counts (as a binary variable) and age to survival time are

interaction model: $h_i(t \mid F, age) = h_0(t)e^{b_1 F + b_2(age_i - \overline{age}) + b_3 F \times (age_i - \overline{age})}$,

additive model: $h_i(t \mid F, age) = h_0(t)e^{b_1 F + b_2(age_i - \overline{age})}$, and

two-sample model: $h_1(t \mid F, age) = h_0(t)e^{b_1 F}$.

The Cox partial likelihood approach yields estimates of the coefficients for these three nested models (Table 9.1) and, to repeat, no parametric assumption about $h_0(t)$ is necessary. However, unlike the parametric estimates (Chapters 7 and 8), the Cox approach does not produce an estimate of the "intercept" term (previously denoted $\hat{b}_0$). Note that the estimate of a specific

Table 9.2. Three hazard ratios calculated from the interaction model for selected ages 25, 30, and 35 years ($CD4 < 700$ versus $CD4 \geq 700$).

Hazard ratios	$age = 25$	$age = 30$	$age = 35$
$\hat{hr}$	0.313	0.448	0.641
$1/\hat{hr}$	3.195	2.232	1.560

coefficient $\hat{b}_j$ changes depending on the other variables in the model. This is a property of statistical models in general and, not surprisingly, the results (for example, p-values) are a combination of the properties of the data and the choice of the model.

The hazard function models include the variable *age* because the two groups compared ("normal" versus "low" CD4-counts) have different age distributions (mean = $\overline{age}_0$ = 29.000 for "low" and mean = $\overline{age}_1$ = 28.032 for "normal" groups) that possibly influence the comparison of survival times between the two CD4-groups. Specifically, an important question becomes: Does the relationship between "low/normal" CD4-count and survival time differ depending on the age of the individuals compared?

Estimates of the hazard ratio for three selected ages (ages = 25, 30, and 35 years—Table 9.2) based on the interaction model describe the influence of an individual's age on the CD4/survival relationship. The expression for the hazard ratio contrasting the influence of the two levels of CD4-exposure on survival time for the same age (denoted a_i) is

$$\hat{hr} = \frac{h_i(t \mid F = 1, age = a_i)}{h_i(t \mid F = 0, age = a_i)} = e^{\hat{b}_1 + \hat{b}_3(a_i - \overline{age})} = e^{-0.904 + 0.072(a_i - \overline{age})}.$$

As required, a hazard ratio estimated from the interaction model depends on the age of the individuals considered. That is, the influence of the binary CD4-variable on the hazard ratio is independent of age only when $b_3 = 0$ (additive model).

The estimated model coefficient $\hat{b}_3$ formally measures the magnitude of the $cd4 \times age$ interaction. That is, it reflects dependence (nonadditivity) between CD4-counts and age. As before, several ways exist to statistically assess the influence of sampling variation on this single estimated regression coefficient (Wald's test, a confidence interval, or a likelihood comparison—Chapter 8). The likelihood comparison is the most general. For the comparison of the interaction model ($b_3 \neq 0$) to the additive model ($b_3 = 0$), the likelihood ratio chi-square statistic is

$$X^2 = -2[\log(L_{b_3=0}) - \log(L_{b_3\neq0})] = -2[-159.301 - (-158.827)] = 0.949.$$

A p-value of $P(X^2 \geq 0.949 \mid b_3 = 0) = 0.330$ indicates that the simpler additive model does not differ greatly from the interaction model. From another perspective, a p-value of 0.330 indicates that chance is a plausible explanation of the observed differences among the hazards ratios ("low" versus "normal" CD4-count) estimated for the three ages 25, 30, and 35 years (Table 9.2). In other words, the likelihood ratio assessment suggests that a single estimated hazard ratio is likely an accurate and certainly a simpler summary of CD4-risk. For the example, $\hat{hr} = e^{-0.898} = 0.407$ from the additive model.

An additive model ($b_3 = 0$) provides a considerable simpler interpretation by separating the influence of the CD4-count on the hazard ratio from the influence of age. The coefficient b_1 measures the influence of the CD4-count regardless of the study subject's age and b_2 measures the influence of age regardless of the study subject's CD4-count. The influence of each variable is "adjusted" for the independent influence of the other.

The comparison of the partial likelihood values calculated from the additive model ($b_2 \neq 0$) and the nested two-sample model ($b_2 = 0$) reflects the independent importance of age in the description of the hazard ratios. The specific likelihood ratio chi-square test statistic

$$X^2 = -2[\log(L_{b_2=0}) - \log(L_{b_2\neq0})]$$
$$= -2[-159.559 - (-159.301)] = 0.515$$

produces the p-value of 0.473. An individual's age does not appear to influence survival time (no evidence of a systematic effect). Unlike the influence of age, the binary measure of CD4-risk has a strong and negative association with the hazard ratio. The estimated hazard ratio is $\hat{hr} = 0.407$ with a p-value = 0.008 in the additive model and $\hat{hr} = 0.414$ with a p-value = 0.006 in the two-sample model. These two estimated hazard ratios are similar because of the lack of influence of a study subject's age on survival time. Thus, the hazard function describing risk for the "normal" group is $2.5(1/0.407)$ times lower than that for the "low" group at any survival time.

The term *confounding* informally means the confusing or mixing of the effects that influence a relationship. When an additive model contains several variables, issues arise concerning the influence of one or more of these variables on a specific variable's relationship to the outcome variable. Untangling confounding influences is a goal of statistical models in general. One of the important features of an additive model is that the confounding associated with a specific variable or variables is simply defined and easily determined.

Before confounding is discussed in the context of a survival model, it is useful to describe the simplest case. Consider again the comparison of two groups (coded $F = 0$ and 1) and a second variable (denoted x—a continuous variable) that may influence the observed difference between groups. A two-sample linear model represented by $y = b_0 + b_1 F + b_2 x$ illustrates the primary issue. For example, cholesterol levels (y) may differ between two groups because of the presence of a binary risk factor (F). However, the level of socioeconomic status ($x =$ personal income) may also differ between the compared groups. Differences associated with levels of income (x) potentially interfere with a clear interpretation of the risk factor's influence on the observed difference between cholesterol levels. A direct measure of this influence x on the risk/outcome relationship (confounding) consists of comparing two additive models, a model with the potential confounding variable x included and the same model with the variable x excluded. In terms of the two-sample example, the model $y = b_0 + b_1 F + b_2 x$ is compared to the model $y = B_0 + B_1 F$. The influence of variable x on the relationship between the binary variable F and the outcome y is directly measured by the difference between the estimates of the coefficients b_1 and B_1. Specifically, the difference $\hat{B}_1 - \hat{b}_1$ measures confounding, where $\hat{B}_1$ represents the estimate of B_1 from the model with the variable x excluded and $\hat{b}_1$ represents the estimate of b_1 from the model with the variable x included. The variable x does not have an important confounding influence when $\hat{B}_1 - \hat{b}_1$ is small and inconsequential. Otherwise, the variable x influences the comparison between groups and the two-variable model accounts for this influence. The estimated coefficient $\hat{b}_1$ is then said to be "adjusted" for the influence of variable x. The quotation marks are a reminder that adjustment depends on a specific definition of confounding, an additive model, and, in this example, a linear influence of x.

For this two-sample linear model, an expression for the confounding influence of x is $\hat{b}_2(\bar{x}_1 - \bar{x}_0)$ or

$$\hat{B}_1 - \hat{b}_1 = \hat{b}_2(\bar{x}_1 - \bar{x}_0),$$

where $\bar{x}_0$ is the estimated mean value of the variable x in one group ($F = 0$) and $\bar{x}_1$ is the estimated mean value in the other group ($F = 1$). The two basic features necessary for a variable to have a confounding effect are evident: the sample mean values must differ between groups ($\bar{x}_0 \neq \bar{x}_1$) and the variable x must be directly related to y ($\hat{b}_2 \neq 0$). In other words, the two groups are not balanced for variable x and variable x is not independently related to the outcome y.

The issue of confounding is much the same for an additive proportional hazards model. The two-sample hazards model (repeated) including a potential confounding variable x is

$$h_i(t \mid F, x) = h_0(t)e^{b_1 F + b_2(x_i - \bar{x})}.$$

The degree of confounding is approximately $\hat{b}_2 (\bar{x}_1 - \bar{x}_0)$ and does not differ in interpretation from the linear model case. When $\hat{b}_2$ or $\bar{x}_1 - \bar{x}_0$ is near zero, no reason exists to include the variable x in the model to adjust for its influence (adjust the coefficient b_1). Conversely, when $\bar{x}_0 \neq \bar{x}_1$ and $\hat{b}_2 \neq 0$ produce a substantial value of $\hat{b}_2 (\bar{x}_1 - \bar{x}_0)$, the variable x influences the comparison between groups ($F = 0$ versus $F = 1$) and usefully contributes to the model.

Comparison of survival times between groups with "high" and "low" CD4-counts from the SFMHS data illustrates the estimation and interpretation of a confounding influence. The question becomes: Does an individual's age influence the difference in the survival times between two groups of men based on their CD4-counts ($n_0 = 41$ men with CD4 < 700 and $n_1 = 31$ men with CD4 ≥ 700)? When age is included in the additive hazards model, the estimated CD4-coefficient b_1 is $\hat{b}_1 = -0.898$. When age is excluded from consideration, the estimated CD4-coefficient B_1 is $\hat{B}_1 = -0.873$. The confounding influence associated with age causes the difference $\hat{B}_1 - \hat{b}_1 = -0.873 - (-0.898) = 0.025$. Incidentally, the approximate value is essentially the same where $\hat{b}_2(\overline{age}_1 - \overline{age}_0) = -0.026(28.032 - 29.000) = 0.025$.

Another measure of confounding influence caused by a specific variable is the percent difference between the estimated coefficients $\hat{B}$ and $\hat{b}$. Specifically, for a variable x_i,

$$\% \text{ change} = 100 \times \frac{\hat{B}_i - \hat{b}_i}{\hat{b}_i}.$$

The percentage change is unitless, so, as with all percentages, values associated with several variables can be compared regardless of their original measurement units. From the example, the percentage change in the $CD4$/survival coefficient associated with age is $100 \times (0.898 - 0.873)/0.898 = 2.8\%$. Although no definitive rule exists, a value in the neighborhood of 3% appears small. Several authors suggest that confounding is not an important issue unless more than a 10% change occurs between the estimated coefficients.

An important question remains: How is the degree of confounding assessed? First, it is important to note that the magnitude of confounding depends on both the choice of measurement units and the measure of association. For the $CD4$ example, the logarithms of the relative hazard functions are compared (regression coefficients). A different magnitude of confounding emerges if the confounding is measured in terms of the change in the hazard ratios. The change in hazard ratios is 0.010 ($e^{-0.873} - e^{-0.989} = 0.010$), yielding a percentage change of 2.5%. Of less importance, when survival time is measured in days rather than months, the degree of confounding also changes. In general, assessment of confounding takes on a subjective character because typically no concrete reasons exist to choose specific measurement units or measure of association.

Not accounting for age, for the example data, decreases the estimate of the coefficient b_1 by 0.025. Whether such a reduction is due to random variation or a systematic effect is not a particularly important question. The observed confounding (the difference between estimates) is a property of the collected data and a choice has to be made to use a model that either includes or excludes the confounding variable. This choice is primarily a subject matter decision, and other than estimating the magnitude of the confounding influence, further statistical analysis is not much help.

Creating a binary variable from a more extensive variable, such as creating a binary $CD4$-variable from the original $CD4$-counts, rarely improves the analysis. Two issues arise. First, there is a loss of statistical power. The loss of power (increased variability of the estimated values) comes from the

Table 9.3. Estimated coefficients from the SFMHS interaction and additive *CD4*/age models for three definitions of low/ normal *CD4*-counts (*p*-values in parentheses).

"Low" *CD4*	Interaction model ($\hat{b}_3$)	Additive model ($\hat{b}_1$)
$CD4 \leq 900$	−0.093 (0.350)	−0.625 (0.162)
$CD4 \leq 700$	0.072 (0.340)	−0.899 (0.008)
$CD4 \leq 500$	0.233 (0.003)	−1.562 (<0.001)

failure to use the collected data efficiently. A variable entered directly into the model as reported takes advantage of all information available. Measured in terms of sample size, a continuous variable can be 40% more efficient than creating a binary variable. Thus, a sample of 60 continuous observations achieves approximately the same statistical power as a sample of 100 binary observations.

A second and more important issue is bias. Results of an analysis typically depend on the way a binary variable is defined. Different definitions frequently lead to different analytic results, introducing a usually unwanted subjectivity into assessment and interpretation. The suspicion arises that the observed influence is due, at least in part, to the artificiality of the created explanatory variable. Table 9.3 illustrates this influence on the estimated regression coefficients b_3 and b_1 from the interaction and additive models (Table 9.1) for three definitions of "low" *CD4*-counts. The estimates and statistical tests differ, sometimes considerably, depending on the definition of "low" *CD4*-counts. Classification of a continuous variable into more than two categories creates a similar potential for bias. However, as the number of categories increases, the bias decreases.

Cox estimation for proportional hazards model: *k*-variable case

Once again, the expression for the *k*-variable proportional hazards model is

$$h_i(t \mid x_{i1}, x_{i2}, \ldots, x_{ik}) = h_0(t) \times c_i = h_0(t) \times e^{\sum b_j x_{ij}}.$$

The Cox semiparametric approach employs the same general form as the parametric multivariable model (Chapter 8) but differs in the estimation procedure. As in the two-sample case, the regression coefficients b_j are estimated without specific information or assumptions about the baseline hazard function $h_0(t)$. That is, the distribution of the sampled survival times is again not relevant to the estimation process. The Cox model/estimation is truly nonparametric, in the sense that the estimated regression coefficients are identical when the observed survival times are replaced by their ranks. Furthermore, the estimates are corrected for the bias incurred from noninformative censored observations.

The previously analyzed HIV-positive study subjects measured for $CD4$-counts, β_2-microglobulin levels, and age along with the multivariable additive proportional hazards model

$$h_i(t \mid cd4, \beta, age) = h_0(t) \times c_i = h_0(t) \times e^{b_1(cd4_i - \overline{cd4}) + b_2(\beta_i - \bar{\beta}) + b_3(age_i - \overline{age})}$$

illustrate the partial likelihood estimation and interpretation of the hazards model coefficients. As always, this additive model does not necessarily produce accurate estimates of the risk/survival relationships and the hazard functions are not necessarily proportional. The assessment of additivity and proportionality remains a critical element of the analysis, even for the Cox distribution-free approach.

A good place to start a survival analysis is with the comparison of models including interaction terms to models with selected interaction terms excluded. The difference in partial log-likelihood values likely indicates the influence of the variables excluded from the model. For the HIV/AIDS data, the model containing three pairwise interaction terms is

$$h_i(t \mid cd4, \beta, age) = h_0(t) \times c_i,$$

where the constant of proportionality c_i (hazard ratio) is defined by

$$
\begin{aligned}
c_i = \exp \Big\{ & b_1(cd4_i - \overline{cd4}) + b_2(\beta_i - \bar{\beta}) + b_3(age_i - \overline{age}) \\
& + b_4(cd4_i - \overline{cd4})(\beta_i - \bar{\beta}) + b_5(cd4_i - \overline{cd4})(age_i - \overline{age}) \\
& + b_6(\beta_i - \bar{\beta})(age_i - \overline{age}) \Big\}.
\end{aligned}
$$

Note that, when the explanatory variables equal their mean values, the hazard function $h_i(t \mid \overline{cd4}, \bar{\beta}, \overline{age}) = h_0(t)$. Table 9.4 displays the partial likelihood estimates of the six regression coefficients necessary to describe the postulated nonadditive hazards model applied to the $n = 72$ HIV-positive SFMHS subjects who are less than 36 years old.

Table 9.4. Estimated regression coefficients based on the Cox proportional hazards model containing three two-way interaction terms from the SFMHS data ($n = 72$, age ≤ 35).

Terms	Coefficients	Estimates	Std. errors	p-values
CD4	b_1	−0.0049	0.0073	—
β	b_2	0.5566	1.5611	—
age	b_3	−0.1454	0.2259	—
CD4 × β	b_4	−0.0009	0.0010	0.311
CD4 × age	b_5	0.0002	0.0002	0.466
β × age	b_6	0.0123	0.0474	0.795

LogLikelihood = −151.280

For example, the three estimated "interaction" coefficients ($\hat{b}_4$, $\hat{b}_5$, and $\hat{b}_6$) give no indication of an important pairwise dependence among the three explanatory variables (p-values 0.311, 0.466, and 0.795, respectively). However, ad hoc comparisons among several estimated "interaction" coefficients are not the best assessment of the joint influences among the model variables. For the example, the three p-values (Table 9.4) likely reflect the absence of interactions, but the p-values do not lend themselves to rigorous inferences or simple interpretations.

The difficulty with a one-at-a-time interpretation of the model-estimated "interaction" coefficients arises from the property that the estimated value of each coefficient depends on the presence or absence of the other interaction terms in the model. For the three estimated interaction coefficients ($\hat{b}_4$, $\hat{b}_5$, and $\hat{b}_6$), a tempting interpretation is that the interactions involving age are unimportant because estimates $\hat{b}_5$ and $\hat{b}_6$ appear to reflect only random influences (not significant). However, if these two terms are eliminated from the model, the estimate $\hat{b}_4$ takes on a different value, primarily because all three estimated coefficients are highly correlated. The coefficient $\hat{b}_4 = -0.001$ with p-value $= 0.31$ becomes −0.003 with p-value $= 0.20$. In general, the direction and magnitude of these kinds of changes are not easily predicted from a series of p-values.

The comparison of likelihood values, however, yields a simply interpreted test-statistic and follows the usual pattern. The chi-square likelihood ratio

Table 9.5. Estimated regression coefficients for the additive Cox proportional hazards model applied to the HIV/AIDS data ($n = 72$, age ≤ 35).

Variables	Coefficients	Estimates	Std. errors	p-values
CD4-counts	b_1	-0.003	0.001	<0.001
β-levels	b_2	0.366	0.162	0.024
age	b_3	-0.017	0.037	0.653
		LogLikelihood $= -152.355$		

test statistic measuring the simultaneous influences of the three interaction terms is

$$X^2 = -2[\log(L_{\text{additive}}) - \log(L_{\text{interaction}})]$$
$$= -2[-152.355 - (-151.280)] = 2.151.$$

The test statistic X^2 is an observation from an approximate chi-square distribution with three degrees of freedom (three deleted parameters—$m = 3$) when only random differences exist between the interaction and additive models. The additive model appears to be an adequate and, as always, a simpler representation of the relationships within the sampled sample HIV data (p-value $= P(X^2 \geq 2.151 \mid b_4 = b_5 = b_6 = 0) = 0.542$). Partial likelihood estimated coefficients for the additive regression model are presented in Table 9.5.

⁎The additive model produces substantial evidence that CD4-counts and β_2-microgolbulin levels systematically and independently influence survival time and essentially no evidence of an influence from the study participant's age (CD4-counts: p-value < 0.001, β-levels: p-value $= 0.024$, and age: p-value $= 0.653$).

Statistically commensurate measures ($\hat{B}_j = \hat{b}_j / S_{b_j}$) of the influence of these three variables are as follows:

$$\text{cd4-counts: } \hat{B}_1 = \frac{-0.00283}{0.00083} = -3.438,$$

$$\beta\text{-levels: } \hat{B}_2 = \frac{0.366}{0.162} = 2.260, \text{ and}$$

$$\text{age: } \hat{B}_3 = \frac{-0.017}{0.037} = -0.449.$$

Table 9.6. The efficiency ratio* of the Weibull and Cox models in terms of the standard errors for the two-sample regression coefficients.

Ratio	$\gamma = 1$	$\gamma = 1.5$	$\gamma = 2$	$\gamma = 3$	$\gamma = 4$
$\lambda_1/\lambda_0 = 1$	1.0	1.5	2.0	3.0	4.0
$\lambda_1/\lambda_0 = 2$	1.0	1.6	2.1	3.2	4.2
$\lambda_1/\lambda_0 = 4$	1.2	1.8	2.4	3.5	4.8

$$* \text{ Efficiency ratio} = \sqrt{\frac{\text{variance}(\hat{b}_{\text{Cox}})}{\text{variance}(\hat{b}_{\text{Weibull}})}}.$$

Comparisons among the three commensurate coefficients ($\hat{B}_j$-values) indicate again the relative roles of the explanatory variables in predicting the time between detection of HIV and a diagnosis of AIDS. The Cox-model comparisons are similar to the parallel comparisons based on the parametric assumption that the survival times are a random sample from Weibull probability distribution (Chapter 8).

The interpretations of the estimates, the tests, and their associated p-values from a Cox estimated additive model are not different from additive models in general. This property and the property that the Cox partial likelihood method does not require detailed assumptions or knowledge about the sampled populations are major reasons this approach is widely used. There is, however, a cost.

When a parametric model is "correct," the model parameters are typically estimated with greater efficiency (smaller variances) [2] and, of course, the hazard and survival functions are simply estimated and more intuitively describe the survival data. The two-sample Weibull hazards model shows that, as expected, the Weibull parametric estimation has greater efficiency (lower variance) than the Cox semiparametric estimation (Weibull: $S_{\hat{b}} = 0.557$ and Cox: $S_{\hat{b}} = 0.592$). The difference in statistical efficiency depends primarily on the Weibull shape parameter γ and increases as it increases. Table 9.6 shows the ratio of the standard errors of the estimated two-sample regression model coefficients for selected values of γ for hazard ratios $\lambda_1/\lambda_0 = 1$, 2, and 4. However, for the range usually encountered in the study of human mortality or disease, only small differences exist in estimation precision.

Table 9.7. Comparison of the estimated standard errors for the Weibull and Cox multivariable proportional hazards models (SFMHS data—$n = 72$ for age ≤ 35).

Variable	Weibull			Cox		
	Estimates	Std. errors	p-value	Estimates	Std. errors	p-value
CD4	−0.002	0.001	0.005	−0.003	0.001	<0.001
β	0.265	0.123	0.031	0.366	0.162	0.024
age	−0.010	0.028	0.726	−0.017	0.037	0.653

A summary of multivariable analyses (Cox versus Weibull—Table 9.7) shows that the standard errors from the Weibull model (repeated from Table 8.7—Chapter 8) are only slightly less than those from the Cox model (repeated from Table 9.5) calculated from the same HIV/AIDS data (SFMHS data—$n = 72$ for age ≤ 35). In fact, for these two approaches, no important differences are observed between the parametric and non-parametric analyses.

Survival function

As might be expected, without a parametric model, the estimation of the hazard and survival functions is not straightforward but not impossible. For the Weibull and exponential survival models, these estimates are easily calculated. The estimated parameters in conjunction with the postulated parametric model produce direct and intuitive estimates of the baseline hazard and survival functions. Estimates without a parametric model are not as simple. The essence of the process is similar to product-limit estimation of a survival function. An assumption that the survival function is approximately constant between complete survival times makes it possible to estimate this constant value. As with the product-limit estimate, these constant and conditional estimates are combined to produce an estimated survival function over the range of the sampled data. Although this approach is simple in principle, the details produce a complex equation and its solution is best left to a computer program. However, for the two-sample model when the survival times are unique (no ties), approximate estimates are easily calculated and suggest the process for more complicated models.

Table 9.8. Estimated survival functions based on SFMHS data (African-Americans, $n = 23$ and $d = 17$).

Time	At-risk	$\hat{S}_0(t)^*$	$\hat{S}_0(t)^{**}$	$\hat{S}_1(t)^{\dagger}$
1	23	0.960	0.957	0.911
4	21	0.920	0.911	0.826
5	20	0.878	0.865	0.743
8	19	0.836	0.820	0.664
13	17	0.790	0.772	0.585
14	16	0.745	0.723	0.512
15	15	0.701	0.675	0.445
16	14	0.657	0.627	0.383
18	13	0.613	0.579	0.328
22	11	0.560	0.526	0.260
23	10	0.507	0.474	0.213
25	9	0.449	0.421	0.162
29	6	0.365	0.351	0.101
30	5	0.285	0.281	0.057
31	4	0.209	0.210	0.028
37	3	0.138	0.140	0.011
80	1	0.040	0.000	0.001

* Cox model computer generated—nonsmoker.
** Approximate—nonsmoker.
† $\hat{S}_0(t) = [\hat{S}_0(t)]^{2.278}$—smoker.

Using again the smoking/survival data for the SFMHS African-American subjects (Chapter 7), the computer estimate and the approximate survival functions illustrate (Table 9.8). Figure 9.3 displays the Cox model computer estimated baseline survival function $\hat{S}_0(t)$ for nonsmokers and the corresponding estimated survival function $\hat{S}_1(t)$ for smokers.

An approximate baseline survival function can be estimated using the expression

$$\hat{S}_0(t) \approx \prod e^{-d_i/n_i} \approx \prod \frac{n_i - d_i}{n_i},$$

where d_i represents the number of deaths among n_i individuals in the ith risk set and $i = 1, 2, \ldots, d =$ number of complete survival times. The second expression is the product-limit estimate (Chapter 4).

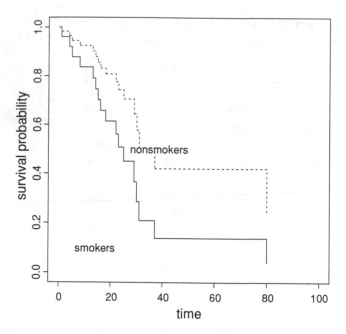

Figure 9.3.   Estimates of the survival functions for $n = 23$ nonsmoking and smoking SFMHS African-American subjects using the Cox proportional hazards model.

Once a baseline survival function is estimated, estimation of survival functions associated with selected levels of the explanatory variables follows the natural pattern, where

$$\hat{S}_i(t \mid x_{i1}, x_{i2}, \ldots, x_{ik}) = [\hat{S}_0(t)]^{\exp(\sum b_j x_{ij})}.$$

Using the two-sample model and smoking data, the estimated survival function for smokers ($F = 1$ and $\hat{b} = -0.823$) is

$$\hat{S}_1(t \mid F = 1) = [\hat{S}_0(t)]^{\exp(\hat{b})} = [\hat{S}_0(t)]^{2.278} \qquad \text{(Table 9.8)}.$$

For example, for $t = 15$ months, $\hat{S}_0(t) = \hat{S}_0(15) = 0.701$(nonsmokers), yielding $\hat{S}_1(t) = \hat{S}_1(15) = (0.701)^{2.228} = 0.445$ (smokers).

Two issues need to be kept in mind in interpreting an estimated survival function:

1. Survival functions estimated from the baseline values [$\hat{S}_0(t)$ produces $\hat{S}_1(t)$] require the hazard functions to be proportional, which is an assumption and not a fact.

2. Also, survival functions estimated from baseline probabilities are frequently extrapolated beyond the limits of the data (a poor idea in most circumstances).

For the example data, the baseline survival function (Table 9.8 and Figure 9.3) is based on the assumption of an additive proportional hazards model and estimated from 17 ($n = 23$ and $d = 17$) survival times ranging from 0 to 80 months. The survival function for the smokers (Figure 9.3) is estimated from a hazards model assumed to be proportional, primarily dominated by data on nonsmokers, and much of the curve represents probabilities extrapolated beyond the observed data for smokers. There are only six smokers and their survival times range from 4 to 23 months.

Goodness-of-fit

The assessment of the Cox estimated proportional hazards model as a summary of the relationships within the survival time data follows the pattern described for parametric models (Chapter 8). The nonparametric estimation process is not an important factor. A primary issue is again the proportionality of the hazard functions. As before, it is essential to verify as well as possible that the influence of explanatory variables does not depend on survival time. When the hazard functions fail to be proportional, the estimated regression coefficients ($\hat{b}_j$) no longer parsimoniously reflect the influence of the explanatory variables. That is, the influence of a single variable is not reflected by a single coefficient.

The SFMHS smoking data (African-American, $n = 23$) provide a simple illustration of exploring goodness-of-fit issues with residual values. As with the Weibull distribution, the Cox–Snell residual values calculated from a two-sample analysis are estimated by

$$r_i = -\log(\hat{S}[t_i \mid F]) = -e^{\hat{b}F} \log(\hat{S}_0[t_i]).$$

The specific r_i-values are given in Table 9.9 based on the computer-estimated survival function $\hat{S}_0(t)$ (Table 9.8—column 3). These residual values are treated much like "data." As before, product-limit estimation produces the probability $\hat{G}(r_i) = P(R \geq r_i)$ associated with each complete survival time (Table 9.10—column 3). For example, the estimated probability of observing

Table 9.9. The Cox–Snell residual values for the two-sample analysis of nonsmokers and smokers among the $n = 23$ African-American SFMHS subjects.

	t_i	$S_0(t_i)$	r_i
1.	1	0.960	0.033
2.	4	0.920	0.154
3.	5	0.878	0.105
4.	8	0.836	0.330
5.	13	0.790	0.190
6.	14	0.745	0.237
7.	15	0.701	0.287
8.	16	0.657	0.339
9.	18	0.613	0.899
10.	22	0.560	0.468
11.	23	0.507	1.248
12.	25	0.449	1.470
13.	29	0.365	0.812
14.	30	0.285	1.012
15.	31	0.209	1.262
16.	37	0.138	1.595
17.	80	0.040	2.595

a residual value greater than $r_5 = 0.237$ is $\hat{G}(r_5) = P(R \geq 0.237) = 0.767$. These probabilities are again a random sample from an exponential distribution with $\lambda = \mu = 1$ (Figure 9.4) when only random differences exist between model and data. The plotted r_i-values (Figure 9.4) appear to correspond to the "unit" exponential distribution.

In addition, the points created by plotting the log-log transformed estimated probability function $\hat{G}(r)$ for each logarithm of the residual value $\log(r)$ randomly deviate from a straight line (intercept $= 0$ and slope $= 1$) when the residual values represent only the random variation associated with a "unit" exponential distribution (Figure 9.5). Specifically, ignoring random variation for the moment, when $G(r) = e^{-r}$, then $\log(-\log[G(r)]) = \log(r)$, a straight line (intercept $= 0$ and slope $= 1$). The least-squares

Table 9.10. Comparison of log(r) and log($-$log[$G(r)$]) to assess the goodness-of-fit of the Cox estimated proportional hazards model from the SFMHS smoking data ($n = 23$).

	r_i	$\hat{G}(r_i)$	log(r_i)	log($-$log[$\hat{G}(r_i)$])
1.	0.033	0.957	−3.423	−3.113
2.	0.105	0.911	−2.254	−2.373
3.	0.154	0.863	−1.872	−1.915
4.	0.190	0.815	−1.661	−1.587
5.	0.237	0.767	−1.438	−1.328
6.	0.287	0.719	−1.248	−1.110
7.	0.330	0.671	−1.109	−0.920
8.	0.339	0.623	−1.081	−0.749
9.	0.468	0.575	−0.759	−0.593
10.	0.812	0.518	−0.208	−0.418
11.	0.899	0.460	−0.107	−0.254
12.	1.012	0.395	0.012	−0.072
13.	1.248	0.329	0.222	0.107
14.	1.262	0.263	0.233	0.289
15.	1.470	0.197	0.385	0.484
16.	1.595	0.132	0.467	0.707
17.	2.595	0.000	0.954	—

estimated straight line in Figure 9.5 has an estimated intercept of 0.005 and slope of 0.966, providing evidence of a "good fit" of the proportional hazards model to the smoking data. Graphically, both plots show no apparent systematic patterns and, therefore, no persuasive reason exists to modify the hazards model. As always, straight lines reflecting fit are easily interpreted and intuitive and provide simple identification of randomness/nonrandomness.

A goodness-of-fit evaluation of the multivariable Cox model applied to the HIV/AIDS data to assess the influences of the CD4-counts, β-levels, and age variables also does not differ in principle from the parametric residual analysis. Figure 9.6 displays the plot of the log-log transformed residual values from the Cox additive model describing the survival experience of the $n = 72$ HIV/AIDS study subjects (Table 9.5). As with the smoking data,

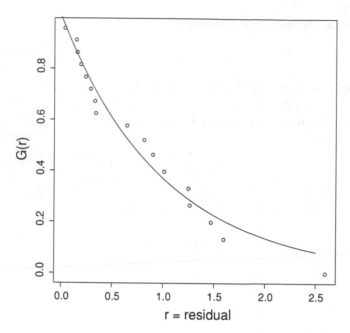

Figure 9.4. Plot of the residual values (r) and their estimated probabilities $\hat{G}(r) = P(R \geq r)$ as well as the "unit" exponential curve e^{-r} from the SFMHS African-American smoking data ($n = 23$).

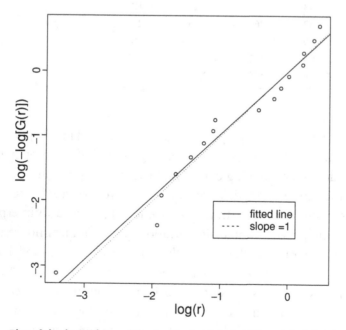

Figure 9.5. Plot of the logarithms of the residual values $\log(r)$ and their log-log transformed probabilities $\log(-\log[\hat{G}(r)])$ from the SFMHS African-American smoking data ($n = 23$).

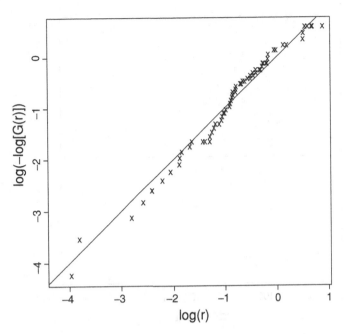

Figure 9.6. Plot of the logarithms of the residual values and their log-log transformed probabilities from the SFMHS HIV/AIDS data ($n = 72$, age ≤ 35).

the plot of the transformed residual values gives no indication of a substantial lack of fit. The intercept and slope of the straight line describing the 72 transformed residual values are intercept $= 0.047$ and slope $= 1.071$. The corresponding values that are expected to occur when the model "fits" are zero and one.

Once residual values are calculated, they can be plotted in a variety of ways to assess randomness. They can be simply plotted in some order (called an *index plot*). For example, the residual values can be plotted by date of entry into the study or by their identification numbers or practically any other potentially useful variable. Residual values can be plotted against survival times, ranks of survival times, explanatory variables in the model, or variables not in the model. They can be transformed and plotted against values of explanatory variables. In short, residual values can be plotted against a long list possibilities with the goal of detecting nonrandom patterns. The process is not governed by a set of rules. These plots are simply a search for nonrandomness. When such a search fails to identify a pattern, confidence

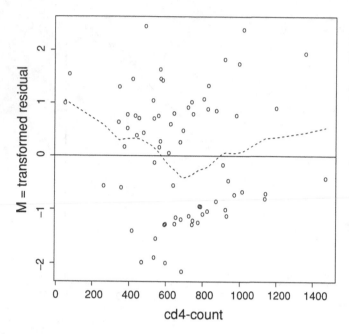

Figure 9.7.   Transformed modified Cox–Snell residual values–CD4-counts.

is increased that the model is at least an adequate representation of the relationships within the collected data.

Among the many choices from plotting residual values, the modified Cox–
Snell residual values (denoted again m_i—Chapter 8) again provide a frequently useful statistical/graphical tool to identify any substantial lack of
fit associated with an explanatory variable. In fact, transformed m_i-values
have an approximately symmetric distribution with mean value zero. Such a
transformation makes the plotted residual values easier to interpret visually.
This somewhat complex transformation is

$$M_i = \text{sign}(r_i)\sqrt{-2(r_i + \delta_i \log[\delta_i - r_i])},$$

where $\text{sign}(r_i) = -1$ for $r_i < 0$ and $+1$ otherwise and $\delta_i = 1$ for complete
observations and $\delta_i = 0$ for censored observations. Figures 9.7, 9.8, and 9.9
are plots of the M_i residual values for the HIV/AIDS CD4-count, the β-level,
and the age explanatory variables from the additive model (Table 9.5). The
plots include a line (smoothed residual values—Chapter 4) summarizing

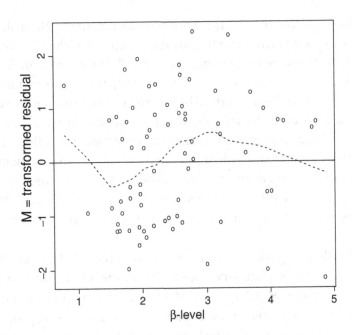

Figure 9.8.   Transformed modified Cox–Snell residual values–β-levels.

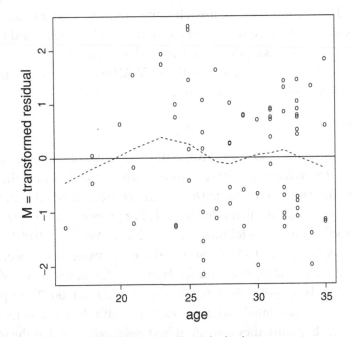

Figure 9.9.   Transformed modified Cox–Snell residual values–age.

the residual pattern (if any) associated with each variable. No striking non-randomness is apparent, with the possible exception of the $CD4$-counts.

To repeat one more time because it is critically important—the influences of the explanatory variables in a survival model (parametric or semiparametric) are summarized and described by hazard functions that are proportional. When the ratio of the hazard functions is constant over the entire range of time considered (proportional!), then and only then, the influence of each explanatory variable is simply summarized. However, an explanatory variable may become less related to survival over time, become more predictive, or change in some other nonproportional way. For example, $CD4$-counts might become a stronger indicator of a diagnosis of AIDS as survival time decreases.

Previously a categorical variable was temporarily created (strata) to explore the linearity of an explanatory risk variable or the interaction between two explanatory variables (Chapter 8). Much the same strategy can be used to explore the assumption that two or more hazard functions are proportional. As before, the data are classified into strata. In the case of assessing the relationship between an explanatory variable and survival time, the strata are based on the survival time itself. It is then a simple matter to compare the strata-specific estimated regression coefficients. When the hazard functions are proportional, these estimates differ by chance alone.

The measured $CD4$-levels from the HIV/AIDS data ($n = 72$ study subjects) and the Cox proportional model hazards model,

$$h_i(t) = h_0(t)e^{b_1 cd4_i},$$

illustrate the comparison of the two estimated values of the regression coefficient b_1. The value ($\hat{b}_l$) estimated from survival times less than the median value is compared to the value ($\hat{b}_u$) estimated from survival times greater than or equal to the median value ($b_l = b_u$? or proportional?). The two estimated coefficients are as follows: for $t < 72.0$ weeks, $\hat{b}_l$ is -0.0037, and for $t \geq 72.0$ weeks, $\hat{b}_u$ is -0.0017 (estimated median value $= 72.0$ weeks). The estimated coefficients appear to differ. However, the approximate 95% confidence intervals are $(-0.0054, -0.0017)$ and $(-0.0046, 0.0012)$, respectively. Because of the substantial variation associated with these estimates ($n = 36$ values in each group), they provide at best weak evidence that the influence of the $CD4$-counts on survival time differs between strata.

Dividing the data into strata is a simple but not a statistically powerful approach to the question: Are the hazard functions proportional? The sample size in each strata is considerably reduced producing in many data sets unstable estimates (high variability). For the example, the estimated coefficients from the stratified analysis of the HIV/AIDS data are each based on 36 observations (33 and 10 complete observations in each group). However, a more sophisticated approach exists that does not require a continuous survival time variable to be subdivided into separate analyses.

A substantially more powerful method is created by incorporating survival time (as reported) directly entered into a hazards model. More formally, an interaction term (variable × time) is added to the model, potentially identifying survival-time-related changes in the influence of an explanatory variable. A simple example of a two-sample hazards model that produces a hazard ratio that is not constant with respect to time (not proportional) is

$$h_1(t) = h_0(t)e^{b_1 F + b_2 \log(t)}.$$

The hazard ratio is

$$\text{hazard ratio} = \frac{h_1(t)}{h_0(t)} = e^{b_1 F + b_2 \log(t)}$$

and obviously depends on survival time except when $b_2 = 0$ (in fact, $\log(\text{hazard ratio}) = b_1 F + b_2 \log(t)$ is a linear function of time).

Continuing the HIV/AIDS data example, an extended proportional hazards model is proposed where

$$h_i(t) = h_0(t)e^{b_1 cd4_i + b_2 [cd4_i \times \log(t_i)]} = h_0(t)e^{[b_1 + b_2 \log(t_i)]cd4_i}.$$

The interaction of $CD4$-counts with survival time ($\log[t_i] \times cd4_i$) is a formal description of time dependency (lack of proportionality). More simply, when b_2 is not zero, the hazard ratio depends on survival time. The estimated regression coefficient $\hat{b}_2$ directly reflects the proportionality assumption and allows a statistical assessment (test) of this conjecture. Because the estimate $\hat{b}_2$ will never be exactly zero when estimated from sampled data, the question becomes: Is there statistical evidence that b_2 is not zero?

The logarithm of survival time $\log(t_i)$ is directly entered into the model primarily for computational reasons. Furthermore, estimation of the "interaction" model coefficients requires special but generally available software. The SAS, SPLUS, STATA, and R statistical analysis systems, for example, allow

Table 9.11. Estimated coefficients from an extended Cox model to explore the possibility of nonproportional hazard functions (SFMHS data, $n = 72$).

Terms	Coefficients	Estimates	Std. errors	p-values
CD4	b_1	−0.0092	0.0028	—
log(time) × CD4	b_2	0.0018	0.0008	0.018
	LogLikelihood = −151.913			

estimation of parameters from a hazards model containing time-dependent interaction terms ($\log(t) \times$ variable).

Using the HIV/AIDS data and the extended proportional hazards model yields the estimated coefficient $\hat{b}_2 = 0.0018$ associated with the CD4/time interaction (Table 9.11). The usual statistical assessment of the estimated b_2-coefficient and, therefore, an assessment of time-dependency associated with the CD4-count is

$$X^2 = \left[\frac{\hat{b}_2 - 0}{S_{\hat{b}_2}} \right]^2 = \left[\frac{0.0018 - 0}{0.0008} \right]^2 = 5.579.$$

The associated chi-square test statistic X^2 (degrees of freedom $= 1$) yields a p-value of $P(X^2 \geq 5.579 \mid b_2 = 0) = 0.018$, suggesting nonproportionality (evidence of an interaction with survival time). In other words, it is likely that the influence of the CD4-count on survival time is not accurately measured by a single coefficient.

The strategy of adding interaction terms to a hazards model to explore possible time dependency (nonproportionality) applies to any number of explanatory variables. The HIV/AIDS data continue to illustrate. The time-dependent (nonadditive) proportional hazards model based on the three explanatory variables, CD4-count, β_2-microgolbulin, and age, becomes

$$h_i(t) = h_0(t) e^{b_1 cd4_i + b_2 \beta_i + b_3 age_i + b_4 [cd4_i \times \log(t_i)] + b_5 [\beta_i \times \log(t_i)] + b_6 [age_i \times \log(t_i)]}.$$

Not unlike the previous nonadditive models, the explanatory variables are no longer independent. Each variable depends on survival time. The six estimated coefficients for this extended Cox proportional hazards model are given in Table 9.12. Both the single-variable model (Table 9.11) and the three-variable model (Table 9.12) indicate a possible lack of proportionality associated with the CD4-counts. The parallel assessments of the β-levels and

Table 9.12. Estimated coefficients from an extended three variable Cox estimated model allowing the possibility of nonproportional hazard functions (SFMHS HIV/AIDS data, $n = 72$ and age ≤ 35).

Terms	Coefficients	Estimates	Std. errors	p-values
CD4	b_1	−0.007	0.003	—
β	b_2	0.420	0.662	—
age	b_3	−0.145	0.143	—
log(time) × CD4	b_4	0.001	0.001	0.084
log(time) × β	b_5	−0.024	0.181	0.894
log(time) × age	b_6	0.038	0.040	0.332
	LogLikehood $= -149.507$			

the study subject's age indicate that a proportional hazards model is probably an adequate representation of their relationship to survival time. However, a comparison between likelihood values from the extended model (Table 9.12) and the additive model (Table 9.5) shows a difference of

$$X^2 = -2[-149.507 - (-152.355)] = 5.697,$$

with a corresponding p-value of 0.127 (degrees of freedom $= 3$).

Stratified analysis

A possible reason for a lack of proportionality among hazard functions is that the baseline hazard functions differ among the levels of another variable. In this case, a proportional hazards model fails to adequately summarize the collected survival data without special modification. A direct and simple strategy consists of conducting a separate analysis within each of several more homogeneous strata. A series of stratified proportional hazards models (one within each stratum) then produces estimates of the regression coefficients unbiased by the stratum variable.

Alternatively, a more efficient hazards model that accounts for differences in baseline hazard functions is

$$h_{ik}(t) = h_k(t)e^{\sum b_j x_{ij}} \qquad k = 1, 2, \ldots, g = \text{number of groups (strata)}.$$

This model postulates a different baseline hazard function for each stratum $[h_1(t), h_2(t), \ldots, h_g(t)]$, whereas the explanatory variable coefficients (b_j) remain the same within all strata. Partial likelihood estimates, nevertheless, produce estimated regression coefficients without defining the g baseline hazard functions. The estimates of the regression coefficients $\hat{b}_j$ can be viewed as pooled estimates of each of g stratum-specific estimates. The estimation of these coefficients requires special estimation techniques but their evaluation and interpretation follow the usual pattern.

Once the regression coefficients are estimated, comparisons of the log–log transformed product-limit estimated survival functions allow a simple graphic assessment of the assumption that the hazard functions are proportional among the strata. That is, within each stratum, log–log transformed estimated survival functions plotted for the values of the logarithm of the survival times deviate randomly from straight and parallel lines (proportional). Thus, the hazard functions differ only because the baseline hazard functions differ. The statistical tests and interpretations of the model-estimated coefficients $(\hat{b}_j)$ are then "free" from the influences of differing baseline hazard functions (the stratum variable).

The $n = 72$ SFMHS participants measured for their $CD4$-counts classified as "hard" drug users (drugs other than marijuana, amyl nitrate, and nitrous oxide) and nonusers illustrate the application of a stratified hazards model. Using Cox proportional hazards partial likelihood estimates modified to account for differing baseline hazard functions within each drug user category gives close to the same estimated $CD4$-coefficient ($\hat{b} = -0.00027$— $S_{\hat{b}} = 0.0008$) that was observed in the previous model (drug-user status ignored—Table 9.5).

Figure 9.10 shows the two stratum-specific product-limit estimated survival functions. Figure 9.11 displays the two log–log transformed product-limit estimated survival functions plotted against the log of the survival time. In symbols, $\log(-\log[\hat{P}_i])$ is plotted against $\log[t_i])$ for each stratum. The least-squares estimated line for users (intercept $= -4.478$ and slope $= 1.055$) appears essentially parallel to the estimated line for nonusers (intercept $= -4.770$ and slope $= 0.943$).

Another useful comparison of the log–log transformed survival functions is achieved by plotting the transformed survival function for nonusers (horizontal axis) against the transformed survival function for users (vertical

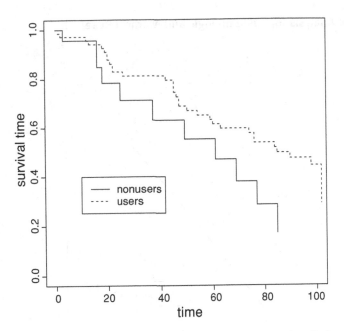

Figure 9.10. Stratified analysis by drug use (presence/absence)–product-limit estimated survival functions for SFMHS data, age less than 36 years, $n = 72$.

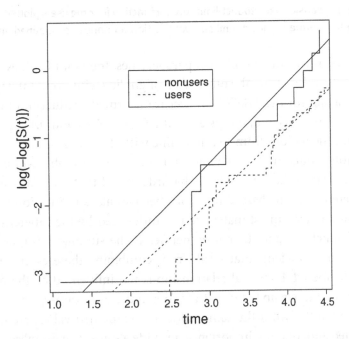

Figure 9.11. Stratified by drug use (presence/absence)–log–log transformed product-limit survival functions for SFMHS data, age less than 36 years, $n = 72$.

217

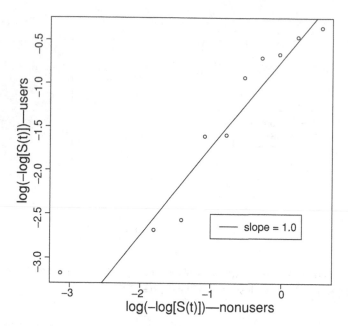

Figure 9.12. The log–log transformed product-limit survival function for drug users plotted against the log–log transformed product-limit survival function for nonusers–stratified analysis.

axis). When two functions form parallel lines, the plot of values from one against the values from the other is a straight line with slope 1.0. Figure 9.12 is such a plot for the AIDS/drug-user transformed product-limit estimated survival functions and, consistent with Figures 9.10 and 9.11, shows only moderate deviations from a straight line with slope = 1.0.

Stratified data in general allow adjustment for a variable without postulating a relationship or model. A hazards model that postulates a number of different baseline hazard functions behaves like stratification in general. The within-stratum estimates are not influenced by the different baseline hazard functions and, therefore, neither is the summary estimates of the regression coefficients that result from combining these estimates. In the example, the *CD4*/survival relationship is not influenced by the drug-user statuses that make up the two strata. When these stratum-specific estimates randomly differ from the same value, the combined values are efficiently estimated and, of more importance, provide a single and useful summary of the relationship to the survival pattern.

Examples of R

Introduction

The R-code and its explanation, presented in the following, are, as they say, "the tip of the iceberg." The statistical language called R is extensive, fully documented online, and described in a number of manuals and books. In fact, there is an entire book completely devoted to the application of R to survival data [11]. Unlike statistical programs (e.g., SAS, STATA, and SPSS), R is a computer language that consists of a large (very large) number of commands. These commands are combined to read data into a computer system, process data into an appropriate form, and create a sequence of functions to statistically analyze the question at hand. This process frequently requires considerable effort, but the payoff is that it is done in an interactive environment that allows complete freedom in choice of techniques and analytic approaches. To use the R computer tools specifically designed for survival analysis, it is necessary to have some knowledge of the R-language in general. Therefore, the following pages assume a basic knowledge of the R-language and are presented more to give a feeling for the R-system than as a detailed description of the computer analysis of survival data.

What follows primarily illustrates the four fundamental R-functions that are at the heart of the computer techniques for the analysis survival data. They are:

1. SURVFIT (product-limit estimation—Chapter 4)
2. SURVREG (parametric model estimation—Chapters 5, 6, 7, and 8)
3. SURDIFF (log-rank test—Chapter 7)
4. COXPH (semiparametric Cox proportional hazards model—Chapter 9).

R-documentation copied directly from the computer "online manual" indicates the general pattern of the R-language. These four fundamental R-functions are obviously a small part of an extensive computer system that is useful for most statistical calculations.

Four R-functions for survival analysis

SURVFIT: Compute a Survival Curve for Censored Data

Description

Computes an estimate of a survival curve for censored data using either the Kaplan–Meier or the Fleming–Harrington method or computes the predicted survivor function for a Cox proportional hazards model.

Usage

```
survfit(formula, data, weights, subset, na.action,
     newdata, individual=F, conf.int=.95, se.fit=T,
     type=c("kaplan-meier","fleming-harrington", "fh2"),
     error=c("greenwood","tsiatis"),
     conf.type=c("log","log-log","plain","none"),
     conf.lower=c("usual", "peto", "modified"))
```

SURVREG: Regression for a Parametric Survival Model

Description

Regression for a parametric survival model. These are all time-transformed location models, with the most useful case being the accelerated failure models that use a log transformation.

Usage

```
survreg(formula=formula(data), data=parent.frame(), weights,
subset,na.action,dist="weibull", init=NULL, scale=0,
control=survreg.control(),parms=NULL,model=FALSE, x=FALSE,
y=TRUE, robust=FALSE, . . .)
```

SURVDIFF: Test Survival Curve Differences

Description

Tests if there is a difference between two or more survival curves using the *G*–rho family of tests, or for a single curve against a known alternative.

Usage

```
survdiff(formula, data, subset, na.action, rho=0)
```

COXPH: Fit Proportional Hazards Regression Model

Description

Fits a Cox proportional hazards regression model. Time-dependent variables, time-dependent strata, multiple events per subject, and other extensions are incorporated using the counting process formulation of Andersen and Gill.

Usage

```
coxph(formula, data=parent.frame(), weights, subset,
    na.action, init, control, method=c("efron","breslow","exact"),
    singular.ok=TRUE, robust=FALSE,
    model=FALSE, x=FALSE, y=TRUE, . . .)
```

The following applied examples using these four R-functions are slightly edited (deleting a few details to focus on the important issues).

Example 1

Estimation for exponential and Weibull survival distributions

#exponential and Weibull survival distributions (Chapters 5 and 6)

library(survival)   #invokes the library of R-functions specific to survival analysis

```
# data
t <- c(1.2,5.0,0.3,3.0,1.3,0.9,7.2,2.3,3.4,
2.7,2.8,1.6,1.1, 1.1,0.7,3.9,1.7,7.3,4.5,
7.5,1.2,0.9,0.6,0.2,2.1,2.1,5.0,4.0,0.8,5.0,
0.5,1.8,3.6,0.1,7.9,4.2,0.1,3.4,0.4,3.6)

cc <- c(0,0,0,1,1,1,0,1,0,1,1,0,1,0,1,0,0,
0,1,0,0,0,0,0,1,0,1,0,0,1,0,0,0,0,1,1,0,1,0,1)
```

n <- length(t)   #number of observations
d <- sum(cc)   #number of complete observations

#exponential distribution estimates

mean <- sum(t)/d
l <- 1/mean
median <- mean*log(2)
round(cbind(n,d,l,mean,median),3)

```
          n    d    l       mean    median
[1,]    40   16   0.150   6.688   4.635
```

#Estimation of the product-limit (Kaplan–Meier) survival curve
fit <- survfit(Surv(t,cc)~1)
summary(fit)

```
Call: survfit(formula=surv(t,cc)~1)

time   n.risk  n.event   survival   std.err   lower   upper
0.7    33      1         0.970      0.0298     0.913   1.000
0.9    31      1         0.938      0.0422     0.859   1.000
1.1    29      1         0.906      0.0517     0.810   1.000
1.3    25      1         0.870      0.0610     0.758   0.998
2.1    21      1         0.828      0.0708     0.701   0.979
2.3    19      1         0.785      0.0794     0.644   0.957
2.7    18      1         0.741      0.0861     0.590   0.931
2.8    17      1         0.698      0.0914     0.540   0.902
3.0    16      1         0.654      0.0955     0.491   0.871
3.4    15      1         0.610      0.0986     0.445   0.838
3.6    13      1         0.563      0.1016     0.396   0.802
4.2    9       1         0.501      0.1079     0.328   0.764
4.5    8       1         0.438      0.1111     0.267   0.720
5.0    7       2         0.313      0.1091     0.158   0.620
```

plot(fit, xlab="time", ylab="survival probability", main="Example data")

#exponential model estimates
fit <- survreg(Surv(t,cc)~1, dist="exponential")
sumary(fit)

```
Call:  survreg (formula=Surv (t,cc)~1, dist="exponential")

              Value Std.   Error   z     p
(Intercept)   1.9          0.25    7.6   2.94e-14

Exonential distribution
Loglik (model)= -46.4   Loglik (intercept only)= -46.4
```

b <- fit$coefficients   #estimated model parameters
L <- −2*fit$loglik[2]   #−2log-likelihood value
mean <- exp(b)
1 <- 1/mean
median <- log(2)/1

```
> round(cbind(n,d,1,mean,median, L), 3)
        n   d   1       mean    median   L
[1,]    40  16  0.150   6.688   4.635    92.808
```

#Weibull distribution model estimates
fit <- survreg(Surv(t,cc)~1, dist="weibull")
summary(fit)

```
Call:  survreg(formula=Surv(t,cc)~1, dist="weibull")
                  Value Std.   Error    z        p
(Intercept)       1.714        0.152   11.30    1.25e-29
Log(scale)       -0.521        0.189   -2.76    5.78e-03

Scale = 0.594

Weibull distribution
Loglik(model) = -43.4      Loglik(intercept only) = -43.4
```

b <- fit$coefficients   #estimated model parameters
L <- −2*fit$loglik[2]   #−2log-likelihood value
s <- fit$scale   #estimated reciprocal of the shape parameter
1 <- exp(−b/s)   #scale parameter
g <- 1/s   #shape parameter
mean <-1^(−1/g)*gamma(1+1/g)
median <-(log(2)/1)^(1/g)

```
round(cbind(n,d,1,g,mean,median,L),3)
                 n    d    1      g      mean    median   L
(Intercept)     40   16   0.056  1.683  4.958   4.667    86.745
```

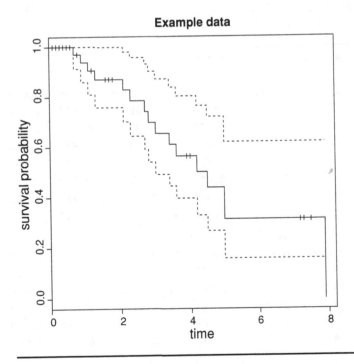

Example data

Example 2

Estimation for exponential and Weibull distribution two-sample models

#Two-sample analysis (Chapter 7)

---

library(survival)   #invokes the library of R-functions specific to survival analysis

```
# data
t <- c(1.2,5.0,0.3,3.0,1.3,0.9,7.2,2.3,3.4,
2.7,2.8,1.6,1.1,1.1,0.7,3.9,1.7,7.3,4.5,
7.5,1.2,0.9,0.6,0.2,2.1,2.1,5.0,4.0,0.8,5.0,
0.5,1.8,3.6,0.1,7.9,4.2,0.1,3.4,0.4,3.6)

cc <- c(0,0,0,1,1,1,0,1,0,1,1,0,1,0,1,0,0,
0,1,0,0,0,0,0,1,0,1,0,0,1,0,0,0,1,1,0,1,0,1)

sex <- c(1,1,0,1,0,1,0,0,1,0,1,0,0,1,1,0,0,0,1,
0,1,0,1,1,0,0,1,0,1,0,0,1,0,1,0,0,1,1,0,1)
```

n <- length(t)   #number of observations
d <- sum(cc)   #number of complete observations

#Product-limit (Kaplan–Meier) estimated survival curve
fit <- survfit(Surv(t,cc)~sex)
summary(fit)

```
Call:  survfit(formula=Surv(t,cc)~ sex)
```

sex=0

time	n.risk	n.event	survival	std.err	lower	upper
1.1	17	1	0.941	0.0571	0.836	1.000
1.3	16	1	0.882	0.0781	0.742	1.000
2.1	13	1	0.814	0.0972	0.645	1.000
2.3	11	1	0.740	0.1131	0.549	0.999
2.7	10	1	0.666	0.1237	0.463	0.959
4.2	6	1	0.555	0.1446	0.333	0.925
5.0	5	1	0.444	0.1525	0.227	0.871

sex=1

time	n.risk	n.event	survival	std.err	lower	upper
0.7	15	1	0.933	0.0644	0.815	1.000
0.9	13	1	0.862	0.0911	0.700	1.000
2.8	8	1	0.754	0.1284	0.534	1.000
3.0	7	1	0.646	0.1485	0.412	1.000
3.4	6	1	0.538	0.1581	0.303	0.957

3.6	4	1	0.404	0.1663	0.180	0.905
4.5	3	1	0.269	0.1561	0.086	0.839
5.0	2	1	0.135	0.1231	0.022	0.808

plot(fit, xlab="time", ylab="survival probabilities", 1ty=1:2, main="Two-sample data")
text(3, 0.4, "Male", cex=1.5)
text(6, 0.5, "Female", cex=1.5)

#exponential distribution estimates (females)
t0 <- t[sex==0]
c0 <- cc[sex==0]
n0 <- length(t0)   #number of complete observations
d0 <- sum(c0)   #number of complete observations
mean0 <- sum(t0)/d0
10 <-1/mean0
median0 <- mean0*log(2)

```
round(cbind(n0,d0,10,mean0,median0),3)
         n0    d0    10      mean0    median0
[1,]     21    8     0.118   8.450    5.857
```

#exponential distribution estimates (males)
t1 <- t[sex==1]
c1 <- cc[sex==1]
n1 <- length(t1)   #number of observations
d1 <- sum(c1)   #number of complete observations
mean1 <- sum(t1)/d1
11 <- 1/mean1
median 1 <-mean1 *log(2)

```
round(cbind(n1,d1,11,mean1,median1),3)
         n1    d1    11      mean1    median1
[1,]     19    8     0.203   4.930    3.414
```

#logrank test (male compared to female survival)
survdiff(Surv(t,cc)~sex)

```
Call:  survdiff(formula=Surv(t,cc)~sex)
```

	N	Observed	Expected	$(O-E)^2/E$	$(O-E)^2/V$
sex=0	21	8	10.21	0.478	1.42
sex=1	19	8	5.79	0.844	1.42

Chisq= 1.4 on 1 degrees of freedom, p= 0.233

```
fit <- coxph(Surv(t,cc)~sex)
summary(fit)


Call:  coxph(formula = Surv(t, cc) ~ sex)
         coef    exp(coef)   se(coef)    z      p
sex     0.619    1.86        0.522     1.18   0.24


        exp(coef)   exp(-coef)   lower.95   upper.95
sex     1.86        0.539        0.667     5.17


Likelihood ratio test= 1.40   on 1 df,   p=0.236
Wald test               = 1.40   on 1 df,   p=0.236
Score (logrank) test = 1.45   on 1 df,   p=0.229
```

```
#exponential model estimates (males/females)
fit <- survreg(Surv(t,cc)~sex, dist="exponential")
summary(fit)


Call:  survreg(formula=Surv(t,cc)~sex, dist="exponential")
                    Value Std.    Error     z        p
(Intercept)         2.13          0.354    6.04    1.58e-09
sex                -0.54          0.500   -1.08    2.80e-01


Exponential distribution
Loglik(model)= -45.8  Loglik(intercept only)= -46.4
     Chisq= 1.15 on 1 degrees of freedom, p= 0.28
```

```
b <- fit$coefficients      #estimated model parameters
L <- -2*fit$loglik[2]      #−2log-likelihood value
mean0 <- exp(b[1])         #estimated scale parameter
ratio <- exp(b[2])         #estimated hazard ratio
mean1 <- mean0*ratio
10 <- 1/mean0
11 <- 10/ratio
median0 <- log(2)/10
median1 <- log(2)/11
```

```
round(cbind(10,11,mean0,mean1,median0,median1,ratio,L), 3)
               10      11     mean0 mean1 median0 median1 ratio L
(Intercept) 0.118 0.203 5.857 3.414 5.857    3.414   0.583 91.656
```

```
#Weibull distribution model estimates
fit <- survreg(Surv(t,cc)~sex, dist="weibull")
summary(fit)
```

```
Call:  survreg(formula=Surv(t,c)~sex, dist="weibull")
                  Value Std.    Error     z        p
(Intercept)       1.891         0.196     9.64     5.30e-22
sex              -0.477         0.273    -1.75     8.08e-02
Log(scale)       -0.605         0.194    -3.12     1.83e-03

Weibull distribution
Loglik(model)= -42  Loglik(intercept only)= -43.4
     Chisq= 2.75 on 1 degrees of freedom, p= 0.097
```

```
s <- fit$scale                #estimated reciprocal of the shape parameter
b <- -fit$coefficients/s      #estimated model parameter
L <- -2*fit$loglik[2]         #-2log-likelihood value
10 <- exp(b[1])
11 <- exp(b[1]+b[2])
g <- 1/s
mean0 <- 10^(-1/g)*gamma(1+1/g)
mean1 <- 11^(-1/g)*gamma(1+1/g)
median0 <- (log(2)/10)^(1/g)
median1 <- (log(2)/11)^(1/g)
```

```
round(cbind(n,d,10,11,g,mean0,mean1,median0,median1,L),3)
              n d 10     11    g     mean0 mean1 median0 median1 L
(Intercept) 40 16 0.0313 0.075 1.832 5.886 3.653 5.423   3.366   83.993
```

```
#plot of male and female survival curves assuming Weibull distributed survival times
plot(fit, xlab="time", lty=1:2, ylab="survival probability", main="Two-sample data")
T <- seq(0,8,0.1)
S0 <- exp(-10*T^g)      #estimated baseline survival function
lines(T, S0, lty=2)
S1 <- exp(-11*T^g)      #estimated survival function
lines(T, S1, lty=3)
text(3,0.4,"Male", cex=1.5)
text(6.5,0.5,"Female", cex=1.5)
label <- c("Product-limit estimates", "Weibull model (female)", "Weibull model (male)")
legend(1, 0.2, label, lty=1:3)
```

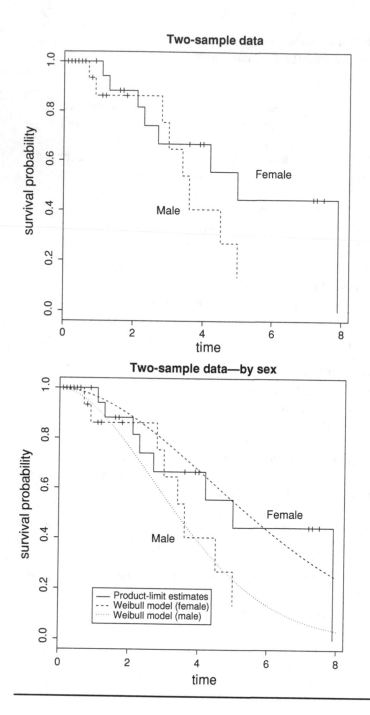

Example 3

Estimation for multivariable proportional hazards models

#Multivariable proportional hazards model (Chapters 8 and 9)

library(survival)   #invokes the library of R-functions specific to survival analysis

```
#data
t <- c(1.2,5.0,0.3,3.0,1.3,0.9,7.2,2.3,3.4,
2.7,2.8,1.6,1.1, 1.1,0.7,3.9,1.7,7.3,4.5,
7.5,1.2,0.9,0.6,0.2,2.1,2.1,5.0, 4.0,0.8,5.0,
0.5,1.8,3.6,0.1,7.9,4.2,0.1,3.4,0.4,3.6)

cc <- c(0,0,0,1,1,1,0,1,0,1,1,0,1,0,1,0,0,
0,1,0,0,0,0,0,1,0,1,0,0,1,0,0,0,0,1,1,0,1,0,1)

sex <- c(1,1,0,1,0,1,0,0,1,0,1,0,0,1,1,0,0,0,1,
0,1,0,1,1,0,0,1,0,1,0,0,1,0,1,0,0,1,1,0,1)

age <- c(40,48,33,42,41,46,53,57,55,48,33,59,44,
52,51,40,33,37,41,41,46,47,46,41,44,53,37,38,
38,47,54,44,42,51,35,43,53,48,37,52)
```

n <- length(t)   #number of complete observations
d <- sum(cc)   #number of complete observations
cbind(n,d)

```
          n       d
[1,]      40      24
```

#Product-limit (Kaplan–Meier) survival curves
fit <- survfit(Surv(t,cc)˜sex)
summary(fit)

Call: survfit(formula = Surv(t, cc) ˜ sex)

sex=0

time	n.risk	n.event	survival	std.err	lower	upper
1.1	17	1	0.941	0.0571	0.836	1.000
1.3	16	1	0.882	0.0781	0.742	1.000
2.1	13	1	0.814	0.0972	0.645	1.000
2.3	11	1	0.740	0.1131	0.549	0.999
2.7	10	1	0.666	0.1237	0.463	0.959
4.2	6	1	0.555	0.1446	0.333	0.925
5.0	5	1	0.444	0.1525	0.227	0.871

```
                        sex=1
   time   n.risk   n.event   survival   std.err   lower   upper
   0.7     15        1         0.933     0.0644    0.815   1.000
   0.9     13        1         0.862     0.0911    0.700   1.000
   2.8      8        1         0.754     0.1284    0.540   1.000
   3.0      7        1         0.646     0.1485    0.412   1.000
   3.4      6        1         0.538     0.1581    0.303   0.957
   3.6      4        1         0.404     0.1663    0.180   0.905
   4.5      3        1         0.269     0.1561    0.086   0.839
   5.0      2        1         0.135     0.1231    0.022   0.808
```

plot(fit, xlab="time", ylab="survival probability", lty=1:2, main="Two-sample data")
text(6, 0.65, "Female", cex=1.5)
text(3, 0.4, "Male", cex=1.5)

```
#age less than 45
fit <- surfit(Surv(t[age<45],cc[age<45])~sex[age<45])
summary(fit)

Call:  survfit(formula=Surv(t[age<45],cc[age<45])~sex[age<45])
```

```
                      sex[age < 45]=0
   time   n.risk   n.event   survival   std.err   lower   upper
   1.1     11        1         0.909     0.0867    0.754   1.000
   1.3     10        1         0.818     0.1163    0.619   1.000
   2.1      8        1         0.716     0.1397    0.488   1.000
   4.2      4        1         0.537     0.1871    0.271   1.000

                      sex [age < 45]=1
   time   n.risk   n.event   survival   std.err   lower   upper
   2.8      4        1         0.75      0.217     0.426   1.000
   3.0      3        1         0.50      0.250     0.188   1.000
   4.5      2        1         0.25      0.217     0.046   1.000
```

plot(fit, xlab="time", ylab="survival probability", lty=1:2, main="Two-sample data")
text(6, 0.9, "age less than 45", cex=1.5)
text(6, 0.65, "Female", cex=1.5)
text(3.5, 0.4, "Male", cex=1.5)

```
#age 45 and older
fit <- survfit(Surv(t[age>=45],cc[age>=45]~sex[age>=45])
summary(fit)
```

```
Call:  survfit(formula=Surv(t [age>=45],cc [age>=45])˜
sex[age>=45])
```

```
                      sex[age >= 45]=0
time   n.risk   n.event   survival   std.err   lower   upper
2.3    4        1         0.75       0.217     0.426   1.000
2.7    3        1         0.50       0.250     0.188   1.000
5.0    2        1         0.25       0.217     0.046   1.000
                      sex[age >= 45]=1
time   n.risk   n.event   survival   std.err   lower   upper
0.7    8        1         0.875      0.117     0.673   1.000
0.9    7        1         0.750      0.153     0.503   1.000
3.4    4        1         0.562      0.199     0.281   1.000
3.6    2        1         0.281      0.222     0.060   1.000
```

plot(fit, xlab="time", ylab="survival probabilitg", 1ty=1:2, main="Two-sample data")
text(4, 0.9, "age 45 and older", cex=1.5)
text(2.0, 0.5, "Female", cex=1.5)
text(4.0, 0.65, "Male", cex=1.5)

#Weibull distribution model—interaction with age?
fit <- survreg(Surv(t,cc)˜(sex+age)^2, dist="weibull")
summary(fit)

```
Call:  survreg(formula=Surv(t,cc)˜(sex+age)^2, dist="weibull")
```

	Value	Std. Error	z	P
(Intercept)	3.1374	1.1986	2.618	0.00886
sex	-2.0919	1.7867	-1.171	0.24167
age	-0.0287	0.0263	-1.092	0.27488
sex:age	0.0370	0.0399	0.928	0.35336
Log(scale)	-0.6306	0.1928	-3.270	0.00108

```
Weibull distribution
Loglik(model) = -41.4  Loglik(intercept only) =
-43.4
      Chisq= 3.96 on 3 degrees of freedom, p= 0.27
```

L <- fit$loglik[2]   #log-likelihood value
fit <- survreg(Surv(t,cc)˜sex+age, dist="weibull")
summary(fit)

```
Call:  survreg (formula = Surv (t, cc)~sex + age, dist =
"weibull")
```

	Value Std.	Error	z	P
(Intercept)	2.4085	0.9016	2.671	0.00755
sex	-0.4514	0.2722	-1.658	0.09728
age	-0.0122	0.0203	-0.598	0.54998
Log(scale)	-0.6202	0.1947	-3.185	0.00145

```
Weibull distribution
Loglik (model)= -41.8  Loglik(intercept only)= -43.4
     Chisq= 3.1 on 2 degrees of freedom, p= 0.21
```

```
L0 <- fit$loglik[2]      #log-likelihood value
X2 <- −2*(L0−L)      #likelihood ratio chi-square statistic
p.value<-1-pchisq(X2,1)
round(cbind(X2,p.value),3)
```

```
        X2      p.value
[1,]   0.864   0.353
```

```
#additive model—parameter estimates
fit <- survreg(Surv(t,cc)~sex+age, dist="weibull")
summary(fit)
```

```
Call:  survreg (formula=Surv(t,cc)~sex+age, dist="weibull")
```

	Value Std.	Error	z	P
(Intercept)	2.4085	0.9016	2.671	0.0076
sex	-0.4514	0.2722	-1.658	0.0973
age	-0.0122	0.0203	-0.598	0.5500
Log(scale)	-0.6202	0.1947	-3.185	0.0015

```
Weibull distribution
Loglik (model)= -41.8  Loglik (intercept only)= -43.4
     Chisq= 3.1 on 2 degrees of freedom, p= 0.21
```

```
s<-fit$scale
b <- −fit$coefficients/s      #estimated model parameters
g<-1/s                        #shape parameter
10<- exp(b[1])                #scale parameter (baseline/females)
11 <- exp(b[1]+b[2])          #scale parameter (males)
median0 <- (−log(0.5)/10)^(1/g)
median1 <- (−log(0.5)/11)^(1/g)
```

```
mean0 <- 10^(−1/g)*gamma(1+1/g)
mean1 <-11^(−1/g)*gamma(1+1/g)

round (cbind (n,d,10,11,g,mean0,mean1,median0,median1),3)
                n   d   10     11     g     mean0 mean1 median0 median1
(Intercept) 40 16 0.011 0.026 1.859 9.873 6.286 9.129   5.812
```

#impact of age? and confounding influence of age?
fit <- survreg(Surv(t,cc)~sex, dist="weibull")
summary(fit)

```
Call: survreg (formula=Surv (t,cc)~sex, dist="weibull")

              Value Std.   Error    z       p
(Intercept)   1.891      0.196    9.64   5.30e-22
sex          -0.477      0.273   -1.75   8.08e-02
Log(scale)   -0.605      0.194   -3.12   1.83e-03

Weibull distribution
Loglik (model)= -42  Loglik(intercept only)= -43.4
     Chisq= 2.75 on 1 degrees of freedom, p= 0.097
```

s <- fit$scale   #estimated reciprocal of the shape parameter
B <- fit$coefficients/s   #estimated parameters
L1 <- fit$loglik[2]   #log-likelihood value
X2 <- −2*(L1−L0)   #likelihood ratio statistic

p.value <- 1−pchisq(X2,1)
round(cbind(X2,p.value),3)

```
        X2      p.value
[1,]   0.347   0.556
```

#confounding

age0 <- mean(age[sex==0])
age1 <- mean(age[sex==1])
B0 <- B[2]   #age removed from the model
b0 <- b[2]   #age included in the model
bias<- B[2]−b[2]

```
round (cbind (age0, age1,B0,b0,bias),3)
          age0     age1      B0       b0      bias
sex     44.095   45.474   0.874    0.839   0.034
```

```
#baseline survival and hazard functions
par(mfrow=c(1,2))
par(pty="s")
T<- 0:10
S <- exp(-10*T^g)        #estimated baseline survival function
plot(T, S, type="1")
title("Survival function")
H <- 10*g*T^(g-1)        #estimated hazard function
plot(T, H, type="1")
title("Hazard function")

#Cox proportional hazards model
f <- coxph(Surv(t,cc)~sex+age)
summary(f)
```

```
Call:  coxph(formula=Surv(t,cc)~sex+age)

  n= 40
        coef    exp(coef)   se(coef)    z       p
sex    0.5834    1.79        0.5251    1.111   0.27
age    0.0288    1.03        0.0403    0.714   0.48

        exp(coef)   exp(-coef)   lower .95    upper .95
sex     1.79        0.558        0.640        5.02
age     1.03        0.972        0.951        1.11

Likelihood ratio test= 1.91   on 2 df,   p=0.385
Wald test             = 1.90   on 2 df,   p=0.387
Score (logrank) test = 1.95   on 2 df,   p=0.377
```

```
par(mfrow=c(1,1))
plot(survfit(f), xlab="time", ylab="survival probability")
title("Baseline survival function—Cox estimated model")
#plots
f0 <- survfit(f)
T <- f0$time
S <- f0$surv        #estimated survival function
plot(T, S, type="s", xlab="time", ylab="survival probability")
S1 <- S^exp(b0)        #b0 from additive model
lines(T, S1, type="s", lty=2)
title("Survival functions—Cox estimated model")
text(2.5, 0.4, "Male", cex=1.5)
text(6, 0.4, "Female", cex=1.5)
```

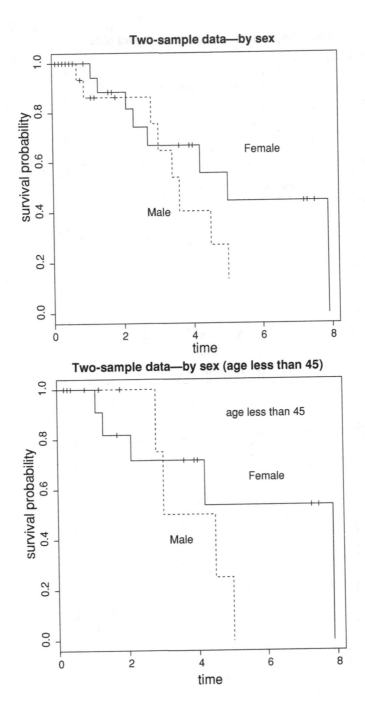

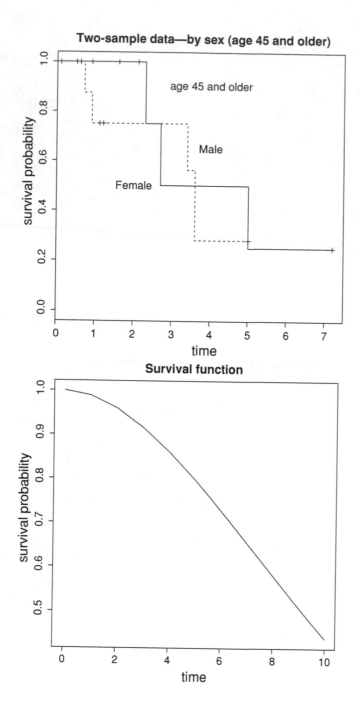

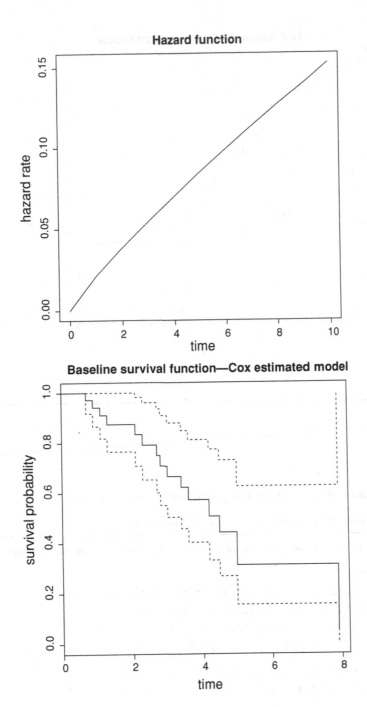

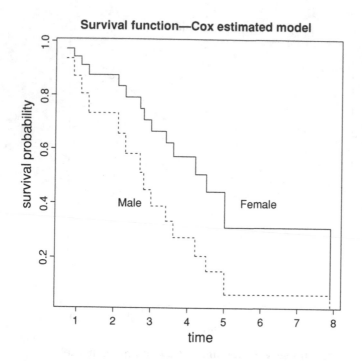

Survival function—Cox estimated model

Example 4

Residual analysis: multivariable proportional hazards models

#Residual values from the Cox approach (Chapter 9)

library(survival)   #invokes the library of R-functions specific to survival analysis

```
# data
t <- c(1.2,5.0,0.3,3.0,1.3,0.9,7.2,2.3,3.4,
2.7,2.8,1.6,1.1, 1.1,0.7,3.9,1.7,7.3,4.5,
7.5,1.2,0.9,0.6,0.2,2.1,2.1,5.0, 4.0,0.8,5.0,
0.5,1.8,3.6,0.1,7.9,4.2,0.1,3.4,0.4,3.6)

cc <- c(0,0,0,1,1,1,0,1,0,1,1,0,1,0,1,0,0,
0,1,0,0,0,0,0,1,0,1,0,0,1,0,0,0,1,1,0,1,0,1)
```

```
sex <- c(1,1,0,1,0,1,0,0,1,0,1,0,0,1,1,0,0,0,1,
0,1,0,1,1,0,0,1,0,1,0,0,1,0,1,0,0,1,1,0,1)

age <- c(40,48,33,42,41,46,53,57,55,48,33,59,44,
52,51,40,33,37,41,41,46,47,46,41,44,53,37,38,
38,47,54,44,42,51,35,43,53,48,37,52)
```

```
n <- length(t)   #number of observations
d <- sum(cc)     #number of complete observations
cbind(n,d)
```

```
         n    d
[1,]    40   24
```

```
#Cox proportional hazards model
fit <- coxph(Surv(t,cc)~sex+age,method='breslow')
summary(fit)
```

```
Call: coxph(formula = Surv(t, cc)~sex + age, method =
"breslow")
```

```
  n= 40
        coef    exp(coef)   se(coef)    z        p
sex    0.5748   1.78        0.5241      1.097    0.27
age    0.0295   1.03        0.0403      0.733    0.46
```

```
        exp(coef)   exp(-coef)   lower .95   upper .95
sex    1.78         0.563        0.636        4.96
age    1.03         0.971        0.952        1.11
```

```
Likelihood ratio test= 1.89   on 2 df,   p=0.390
Wald test             = 1.87   on 2 df,   p=0.392
Score (logrank) test = 1.93    on 2 df,   p=0.381
```

```
r <- f$residual     #modified residuals
rr <- cc-r          #recovering the Cox–Snell residual values from the modified values
fit <- survfit(Surv(rr,cc)~1)
S <- fit$surv       #estimated baseline survival function
T <- fit$time
```

```
#plot residual values
plot(T, S, xlim=range(T), ylim=c(0,1), xlab="time", ylab="residual values")
```

```
title("Residual plot: residual values (exponential?)")
t0 <- seq(0,max(T),0.05)
S0 <- exp(-t0)   #estimated baseline survival function
lines(t0,S0)
```

```
#transformed
plot(log(T), log(-log(S)), xlab="log(time)", ylab="log(-log[residual values])")
abline(0,1)
title("Residual plot: transformed residual values (straight line?)")
```

```
#binomial test of distribution censored values: males versus females
```

```
n0 <- length(cc[sex=0])
n <- length(cc)
binom.test(n0, n, p=0.5)
```

```
        Exact binomial test

data:  n0 and n
number of successes = 21, number of trials = 40, p-value =
0.8746 alternative hypothesis: true probability of success is
not equal to 0.5
95 percent confidence interval:
  0.3612801 0.6848803
sample estimates:
probability of success
        0.525
```

```
#transformed modified Cox–Snell residuals by age
```

```
m <- sign(r)*sqrt(-2*(r+cc*log(cc-r)))
plot(age, m, xlab="age", ylab="modified residual values")
abline(h=0)
title("Modified residual values by age—example data")
```

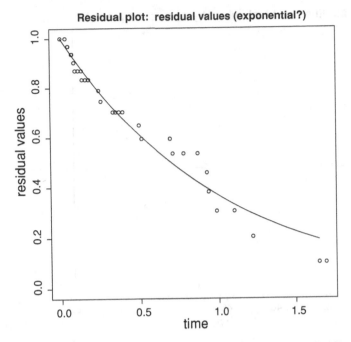

Residual plot: residual values (exponential?)

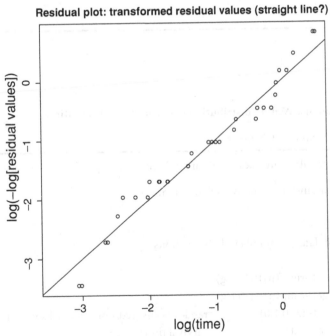

Residual plot: transformed residual values (straight line?)

Modified residual values by age—example data

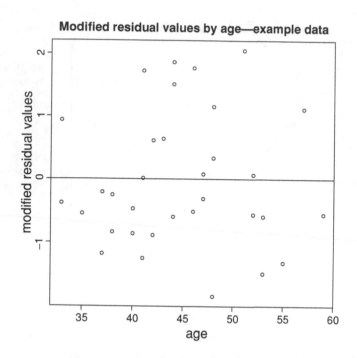

Example 5

Simulation of a Weibull distribution with censored observations

#simulate.weibull.r (Chapter 8)

```
library(survival)   #invokes the library of R-functions specific to survival analysis

#sets the parameters of the Weibull distribution
l<- 0.04
g<- 2.5
#simulated data: n = number of observations
n<-100
t1 <- (−log(1−runif(n))/1)^(1/g)
t2 <- (−log(1−runif(n))/1)^(1/g)
cc <- ifelse(t1−t2<0,1,0)        #random censored/noncensored survival times
t <- ifelse(cc==1,t1,t2)         #survival times
```

#Estimate the Weibull parameters from the simulated data
fit <- survreg(Surv(t,cc)˜1,dist="weibull")
summary(fit)

```
Call: survreg(formula = Surv(t, cc) ˜ 1, dist = "weibull")
                  Value Std.       Error       z        p
(Intercept)       1.160           0.0489     23.7     2.85e-12
Log(scale)       -0.946           0.0986     -9.6     8.33e-22

Scale= 0.388

Weibull distribution
Loglik (model) = -115.9    Loglik(intercept only) = -115.9
```

L <- exp(−fit$coefficients/s)
G <- 1/fit$scale

```
round(cbind(1,L,g,G),4)
                1      L       g      G
(Intercept)    0.04   0.0374  2.5   2.568
```

#plot the estimated survival function
o <- order(t)
c0 <- cc[o]
t0 <- t[o]
S0 <- exp(−L*t0ˆG)
plot(t0, S0, type="1", xlab="time", ylab="Survival probability")

#Calculate residual values and plot probability distribution
r< −log(S0)
f <- survfit(Surv(r,c0)˜1)
r0 <- f$surv
plot(r,r0, pch="o", xlab="residual", ylab="probability")
lines(r, exp(−r))

#Transform the residual values and plot
plot(log(r), log(−log(r0)), pch="o", xlab="log(residual)", ylab="transformed residu-
als") abline(0,1)

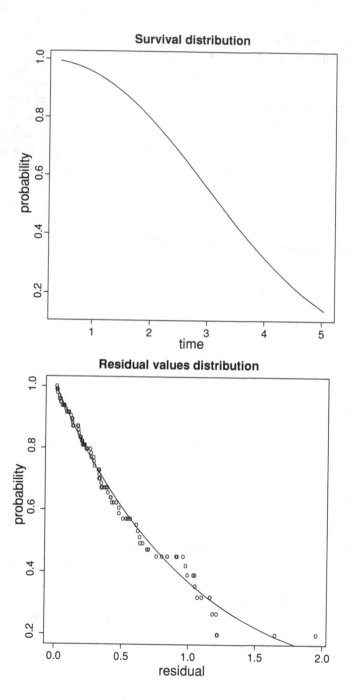

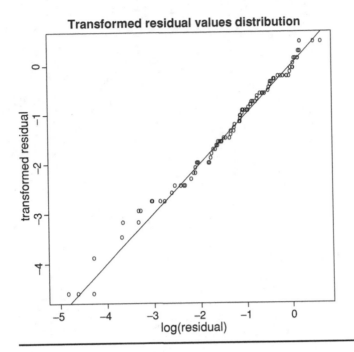

Data

Table A.1. AIDS data—23 African-American men from the San Francisco Men's Health Study.

	Time	Status	Smoker
1	1	1	1
2	2	0	1
3	4	1	0
4	5	1	1
5	8	1	0
6	12	0	1
7	13	1	1
8	14	1	1
9	15	1	1
10	16	1	1
11	18	1	0
12	21	0	0
13	22	1	1
14	23	1	0
15	25	1	0
16	26	0	1
17	27	0	1
18	29	1	1
19	30	1	1
20	31	1	1
21	37	1	1
22	42	0	1
23	80	1	1

Table A.2. AIDS data—174 white men from the San Francisco Men's Health Study.

	t	cc	smk		t	cc	smk		t	cc	smk		t	cc	smk
1	1	1	1	45	9	1	1	89	18	1	1	133	25	1	1
2	1	1	1	46	9	1	1	90	18	1	1	134	26	1	1
3	1	1	0	47	9	1	0	91	18	1	0	135	26	1	0
4	1	1	1	48	9	1	1	92	19	1	1	136	26	1	1
5	1	1	1	49	9	1	1	93	19	1	1	137	26	1	1
6	1	1	1	50	9	1	1	94	19	1	1	138	27	1	1
7	2	1	0	51	10	1	0	95	19	1	0	139	27	1	0
8	2	1	0	52	10	1	0	96	19	1	0	140	27	0	0
9	2	1	0	53	10	1	0	97	19	1	0	141	27	1	0
10	2	1	1	54	10	1	1	98	19	1	1	142	28	1	1
11	2	1	0	55	10	1	0	99	20	1	0	143	28	1	0
12	2	1	0	56	10	1	0	100	20	1	0	144	28	1	0
13	2	1	1	57	10	1	1	101	20	1	1	145	29	1	1
14	3	1	0	58	11	1	0	102	20	1	0	146	29	1	0
15	4	1	0	59	11	1	0	103	20	0	0	147	30	1	0
16	4	1	1	60	11	1	1	104	20	1	1	148	31	1	1
17	4	1	0	61	12	1	0	105	21	1	0	149	31	1	0
18	4	1	1	62	12	1	1	106	21	1	1	150	33	1	1
19	5	1	1	63	12	1	1	107	21	1	1	151	34	0	1
20	5	1	1	64	12	0	1	108	21	1	1	152	35	1	1
21	5	1	1	65	13	1	1	109	21	1	1	153	35	1	1
22	6	1	1	66	13	1	1	110	21	1	1	154	38	1	1
23	6	1	1	67	14	1	1	111	21	1	1	155	38	1	1
24	7	1	1	68	14	1	1	112	21	1	1	156	42	1	1
25	7	1	0	69	14	1	0	113	22	1	0	157	42	0	0
26	7	1	1	70	15	1	1	114	22	1	1	158	44	1	1
27	7	1	1	71	15	1	1	115	22	1	1	159	44	1	1
28	7	1	0	72	15	1	0	116	23	1	0	160	44	0	0
29	7	1	1	73	15	1	1	117	23	1	1	161	48	1	1
30	7	1	0	74	15	1	0	118	23	1	0	162	52	1	0
31	8	1	1	75	15	1	1	119	23	1	1	163	55	0	1
32	8	1	0	76	15	1	0	120	23	0	0	164	57	1	0
33	8	1	0	77	15	1	0	121	23	1	0	165	58	0	0
34	8	1	1	78	15	1	1	122	24	1	1	166	61	0	1
35	8	1	1	79	16	1	1	123	24	0	1	167	61	0	1
36	8	1	0	80	16	1	0	124	24	1	0	168	61	1	0
37	8	1	1	81	16	1	1	125	24	1	1	169	66	0	1
38	8	1	0	82	16	1	0	126	24	1	0	170	67	0	0
39	8	1	1	83	16	1	1	127	24	1	1	171	70	0	1
40	8	1	1	84	16	1	1	128	24	1	1	172	73	0	1
41	8	1	1	85	17	1	1	129	25	1	1	173	95	0	1
42	8	1	1	86	17	1	1	130	25	1	1	174	108	0	1
43	8	1	1	87	17	1	1	131	25	0	1	—	—	—	—
44	8	1	1	88	17	1	1	132	25	1	1	—	—	—	—

Note: t = time, cc = censored/complete, and smk = nonsmoker/smoker.

Table A.3. HIV/AIDS data—88 white men from the San Francisco Men's Health Study.

	Time	Status	cd4	β	Age		Time	Status	cd4	β	Age
1	1	1	485	2.79	25	45	69	1	615	2.79	18
2	2	1	74	2.74	21	46	70	0	415	2.06	34
3	3	1	49	3.90	24	47	74	1	827	2.20	31
4	5	0	263	5.51	44	48	75	1	598	2.55	38
5	6	0	908	2.18	21	49	76	1	568	1.84	28
6	12	1	1020	3.36	25	50	76	1	692	2.08	26
7	13	1	346	3.70	32	51	77	1	536	2.71	31
8	16	1	341	4.67	20	52	80	0	724	2.61	32
9	16	1	567	2.59	27	53	84	0	1013	1.81	30
10	18	1	389	4.13	29	54	84	1	747	1.49	33
11	18	1	431	4.73	33	55	85	1	640	4.02	28
12	19	1	570	2.21	25	56	85	1	356	1.97	29
13	20	1	581	2.12	32	57	90	1	676	2.03	28
14	20	1	417	0.79	33	58	95	0	788	1.68	33
15	21	1	918	2.58	35	59	96	0	681	2.17	35
16	22	1	296	3.03	41	60	96	0	493	2.97	47
17	22	1	528	2.62	33	61	97	0	970	1.66	33
18	22	1	451	3.19	31	62	97	0	597	1.64	29
19	23	0	341	2.61	36	63	97	0	648	1.79	24
20	25	1	389	3.22	33	64	97	0	872	1.52	28
21	26	1	373	3.62	26	65	97	0	825	2.40	32
22	37	0	937	1.81	18	66	98	1	873	1.60	32
23	37	1	830	3.15	34	67	98	0	534	3.00	26
24	42	1	544	1.96	39	68	98	0	742	2.02	16
25	42	1	996	1.74	23	69	98	0	922	2.53	26
26	45	1	261	3.95	32	70	98	0	928	3.22	27
27	45	1	533	2.40	30	71	98	0	1140	2.55	32
28	45	1	439	2.77	31	72	100	0	592	1.60	32
29	46	1	643	2.67	29	73	100	0	745	1.95	21
30	47	1	805	2.39	26	74	100	0	1468	1.97	25
31	47	1	719	2.57	31	75	100	0	598	3.95	30
32	48	1	297	2.45	36	76	100	0	654	1.61	35
33	49	0	764	1.79	38	77	101	0	406	2.98	36
34	49	1	558	1.76	31	78	101	0	799	2.34	33
35	50	1	738	1.86	28	79	101	0	782	1.15	27
36	54	1	1359	1.94	23	80	101	1	800	2.35	36
37	56	0	478	2.11	38	81	101	0	774	2.46	24
38	57	1	906	3.39	38	82	101	0	543	1.95	26
39	59	1	981	4.22	24	83	101	0	685	4.86	26
40	60	1	476	1.69	33	84	101	0	897	1.71	38
41	61	1	562	2.66	25	85	102	0	467	1.78	34
42	62	1	677	2.83	37	86	102	0	413	2.46	40
43	63	1	625	2.12	35	87	102	1	1199	2.67	33
44	65	1	646	2.94	41	88	102	0	1137	1.98	33

Table A.4. Hypothetical data ($n = 40$).

	Time	Status	Sex	Age
1	1.2	0	1	40
2	5.0	0	1	48
3	0.3	0	0	33
4	3.0	1	1	42
5	1.3	1	0	41
6	0.9	1	1	46
7	7.2	0	0	53
8	2.3	1	0	57
9	3.4	0	1	55
10	2.7	1	0	48
11	2.8	1	1	33
12	1.6	0	0	59
13	1.1	1	0	44
14	1.1	0	1	52
15	0.7	1	1	51
16	3.9	0	0	40
17	1.7	0	0	33
18	7.3	0	0	37
19	4.5	1	1	41
20	7.5	0	0	41
21	1.2	0	1	46
22	0.9	0	0	47
23	0.6	0	1	46
24	0.2	0	1	41
25	2.1	1	0	44
26	2.1	0	0	53
27	5.0	1	1	37
28	4.0	0	0	38
29	0.8	0	1	38
30	5.0	1	0	47
31	0.5	0	0	54
32	1.8	0	1	44
33	3.6	0	0	42
34	0.1	0	1	51
35	7.9	1	0	35
36	4.2	1	0	43
37	0.1	0	1	53
38	3.4	1	1	48
39	0.4	0	0	37
40	3.6	1	1	52

Problem set

Chapter 1. Rates and their properties

1. An expression for an average approximate rate from time t to time $t + \delta$ is

$$R_t = \frac{S(t) - S(t + \delta)}{\delta(\frac{1}{2})[S(t) + S(t + \delta)]}.$$

Suppose $S(t) = e^{-\lambda t} = e^{-0.015t}$.

i. Find R_{20} and R_{60} for $\delta = 10.0$.

ii. Find R_{20} and R_{60} for $\delta = 1.0$.

iii. Find R_{20} and R_{60} for $\delta = 0.1$.

iv. Show that $(\delta/2)[S(t) + S(t + \delta)] = \delta[S(t + \delta) + \frac{1}{2}d(t)] = \delta[S(t) - \frac{1}{2}d(t)]$, where $S(t) = P(T \geq t)$ and $d(t) = S(t) - S(t + \delta)$.

v. Show that, when $S(t) = (b - t)/(b - a)$, $R = $ average mortality rate $= 1/[b - \frac{1}{2}(t_1 + t_2)]$.

2. Binomial/normal based 95% confidence interval:
   If $X = 4$ outcomes occur among $N = 200$ possibilities, then the proportion $\hat{q} = X/N = 4/200 = 0.02$ is an estimate of the underlying binomial probability q.
   i. Construct an approximate-normal-based 95% confidence interval from this estimate, where

   lower limit $= \hat{q} - 1.960\sqrt{\text{variance}(\hat{q})} = $ _____?

   upper limit $= \hat{q} + 1.960\sqrt{\text{variance}(\hat{q})} = $ _____?

Using the same date, apply a logistic transformation to created a 95% confidence interval.

ii. Construct an approximate-normal-based 95% confidence interval using the logistic transformed estimate $\hat{q}$ (denoted $\hat{l}$), where

$$\text{lower limit} = \hat{l} + 1.960\sqrt{\text{variance}(\hat{l})} = \underline{\hspace{2cm}}?$$

$$\text{upper limit} = \hat{l} - 1.960\sqrt{\text{variance}(\hat{l})} = \underline{\hspace{2cm}}?$$

iii. Now, use this result to estimate a 95% confidence interval based on $\hat{q}$

$$\text{lower limit} = \underline{\hspace{2cm}}?$$

$$\text{upper limit} = \underline{\hspace{2cm}}?$$

Note that the exact bounds are (0.0055, 0.0504).

3. Use the logistic transformation to investigate the differences in prostatic cancer rates between married and single men, for both races and for whites and blacks seqarately.

Prostatic cancer by race, age, and marital status: cases (1969–1971)

	White			Black		
Age	35–44	45–54	55–64	35–44	45–54	55–64
Single	3	11	18	1	4	2
Married	16	100	187	2	11	29

Prostatic cancer by race, age, and marital status: populations

	White			Black		
Age	35–44	45–54	55–64	35–44	45–54	55–64
Single	74,457	61,665	46,009	12,374	7,569	4,492
Married	923,669	929,227	670,266	79,874	75,641	54,154

Prostatic cancer by race, age, and marital status: logistic transformed values

	White			Black		
Age	35–44	45–54	55–64	35–44	45–54	55–64
Single	——	——	——	——	——	——
Married	——	——	——	——	——	——

Prostatic cancer by race, age, and marital status: variances

Age	White			Black		
	35–44	45–54	55–64	35–44	45–54	55–64
Single	——	——	——	——	——	——
Married	——	——	——	——	——	——

A contrast is defined, as $\hat{c} = \sum a_i \hat{l}_i$ when the $\hat{l}_i$-values are independent logistic transformed probabilities and $\sum a_i = 0$. The variance of a contrast is variance$(\hat{c}) = \sum a_i^2$ variance$(\hat{l}_i)$. For example, a contrast $\hat{c} = l_1 + l_2 - (\hat{l}_3 + \hat{l}_4)$ has an estimated variance of variance$(\hat{c}) =$ variance$(\hat{l}_1) +$ variance$(\hat{l}_2) +$ variance$(\hat{l}_3) +$ variance$(\hat{l}_4)$.

Test the three conjectures with the above data and detemine the associated p-value:

    i. Single does not differ from married in prostatic cancer risk—all men

    ii. Single does not differ from married in prostatic cancer risk—white only

    iii. Single does not differ from married in prostatic cancer risk—black only

4. Heterogeneity of the probabilities q_i:

    Two basic rules of variance are (1) $\sum$ variance$(x_i) =$ variance$(\sum x_i)$ when the x_i-values are uncorrelated and (2) for the constant a, variance$(ax_i) = a^2$variance(x_i), always.

    i. Show that when $\hat{q} = \sum n_i \hat{q}_i / n$, then the variance$(\hat{q})$ is $\sum n_i \hat{q}_i (1 - \hat{q}_i)/n^2$, where $\hat{q}_i = x_i/n_i$, variance$(x_i) = n_i \hat{q}_i (1 - \hat{q}_i)$ and $n = \sum n_i$.

    ii. Create an example using a set of hypothetical numbers x_i and n_i ($q_i = x_i/n_i$) that demonstrate that bias $= \hat{n}q(1 - \hat{q}) - \sum n_i \hat{q}_i (1 - \hat{q}_i) = \sum n_i (\hat{q}_i - \hat{q})^2$, where $\hat{q} = \sum x_i/n$.

Chapter 2. Life tables

1. In general, $L_x = (l_x - d_x) + \bar{a}_x d_x$ but the last open-ended interval is a special case and $L_{x+} = (l_{x+} - d_{x+}) + \bar{a}_{x+} d_{x+}$; thus

$$l_{x+} - d_{x+} = \underline{\hspace{1cm}}? \quad \text{and} \quad \bar{a}_{x+} = \underline{\hspace{1cm}}? \quad \text{making } L_{x+} = \frac{l_{x+}}{R_{x+}},$$

where R_{x+} is the current mortality rate for individuals beyond age x.

Verify that $L_{90+} = 51{,}572$ (Chapter 2, Table 2–1).

2. Data—survival times ($n = 30$):

45.5, 70.5, 26.8, 1.7^+, 21.5^+, 78.1^+, 11.6, 38.5^+, 0.5, 37.5, 29.1^+, 31.3^+, 31.3, 50.0, 2.0, 45.1, 49.3, 23.5^+, 9.1, 10.3^+, 65.9, 12.7^+, 0.3, 57.7^+, 29.9, 57.7^+, 24.6^+, 15.1^+, 18.0, 15.9, where $n - d = 13$ censored and $d = 17$ complete observations.

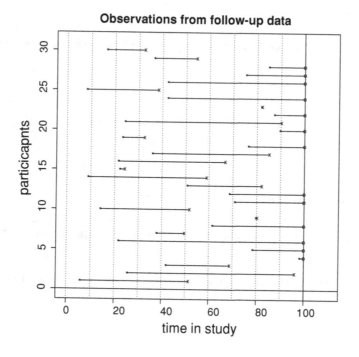

Observations from follow-up data

Fill in the following life table:

i	Interval	Deaths	Censored	l_i	l'_i	$\hat{q}_i$	$\hat{p}_i$	$\hat{P}_k$
1	0–10	——	——	——	——	——	——	——
2	10–20	——	——	——	——	——	——	——
3	20–30	——	——	——	——	——	——	——
4	30–40	——	——	——	——	——	——	——
5	40–50	——	——	——	——	——	——	——
6	50–60	——	——	——	——	——	——	——
7	60–70	——	——	——	——	——	——	——
8	70–80	——	——	——	——	——	——	——
9	80–90	——	——	——	——	——	——	——
10	90–100	——	——	——	——	——	——	——

3. The survival probabilities from John Graunt's original life table are:

Age	0	10	20	30	40	50	60	70	80	90
$\hat{P}_x$	1.0	0.54	0.34	0.21	0.14	0.08	0.05	0.02	0.01	0.00

Plot the survival curve from the 17th-century data.

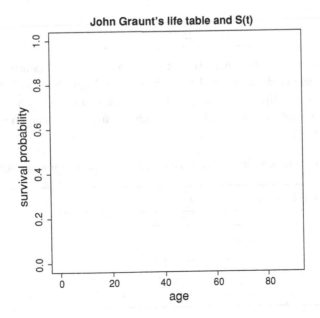

4. Calculate the crude mortality rate ($\hat{\lambda}$) from Table 2.3 (Chapter 2). Estimate and display the exponential survival curve $\hat{S}(t)$. That is, for t from 0 to 90, plot $\hat{S}(t) = e^{-\hat{\lambda}t}$ on the previous set of axes. Interpretation?

5. Calculate a three-year survival probability from the following annual cohorts of data; include the standard error and a 95% confidence interval ($x =$ time survived from diagnosis):

Year	x to $x + 1$	l_x	d_x	u_x	w_x
1996	0–1	9	3	1	—
	1–2	5	1	0	—
	2–3	4	1	0	—
	3–4	3	1	0	—
	4–5	2	0	0	2
1997	0–1	17	10	0	—
	1–2	7	0	0	—
	2–3	7	1	0	—
	3–4	6	2	2	2
1998	0–1	29	10	0	—
	1–2	19	11	0	—
	2–3	6	0	0	6
1999	0–1	34	10	0	—
	1–2	24	3	3	18
2000	0–1	50	25	1	24

i. First, combine the cohort data into a summary table and estimate the effective number of individuals at risk for each year after diagnosis (x to $x + 1$), namely l'_x. Note: $l_x =$ alive a time x, $d_x =$ died in the interval x to $x + 1$, $u_x =$ missing in the interval x to $x + 1$, and $w_x =$ withdrawn from consideration in the interval x to $x + 1$.

Calculation of a survival probability from cohort data: summary data

x to $x + 1$	l_x	d_x	u_x	w_x
0–1	_____	_____	_____	_____
1–2	_____	_____	_____	_____
2–3	_____	_____	_____	_____
3–4	_____	_____	_____	_____
4–5	_____	_____	_____	_____

Calculation of a survival probability from cohort data: computations

x to $x+1$	l'_x	d_x	$\hat{q}_x$	$\hat{p}_x$	$\hat{P}_k$
0–1	———	———	———	———	———
1–2	———	———	———	———	———
2–3	———	———	———	———	———
3–4	———	———	———	———	———
4–5	———	———	———	———	———

ii. Estimate the three-year survival probability $\hat{P}_3 = $ _____?

iii. Construct an approximate 95% confidence interval based on this estimate:

   lower bound = _____? and upper bound = _____?

iv. The estimated variance of the distribution of the estimate $\hat{P}_k$ is
   variance($\hat{P}_k$) $= \hat{P}_k^2 \sum \hat{q}_i/(\hat{p}_i n_i)$ (Greenwood's expression for the variance of $\hat{P}_k$
   for $i = 1, 2, \ldots, k$).

Chapter 3. Two especially useful estimation tools

1. The probabilities from the Poisson probability distribution are given by the expression

$$P(X = k) = \frac{e^{-\lambda}\lambda^k}{k!}.$$

Consider the following random sample from a Poisson probability distribution
($n = 10$):

data (x_i): 0, 2, 4, 0, 1, 2, 3, 0, 0, and 1.

i. Construct the likelihood expression for these data based on the Poisson probability
distribution function. Specifically, for any value of λ, the likelihood value L is
$L = \prod_{i=1}^{10} P(X = x_i)$.

ii. Evaluate the log-likelihood values:

   for $\lambda = 1.0$ $\log(L) = $ _____?

   for $\lambda = 1.2$ $\log(L) = $ _____?

   for $\lambda = 1.4$ $\log(L) = $ _____?

   for $\lambda = 1.6$ $\log(L) = $ _____?

iii. Compute the mean value: $\bar{x} = $_____?

iv. Evaluate the log-liklihood values:

for $\lambda = \bar{x}$ $\log (L) = $_____?

for $\lambda = \bar{x} - 0.01$ $\log (L) = $_____?

for $\lambda = \bar{x} + 0.01$ $\log (L) = $_____?

With a bit of calculus, it can be shown that the maximum likelihood estimate of λ is in general $\bar{x}(\hat{\lambda} = \bar{x})$.

v. Test the conjecture that $\lambda = 2$:

Calculate log-likelihood for $\lambda = \bar{x}$; then $-2 \log (L) = $_____?

Calculate log-likelihood for $\lambda_0 = 2$; then $-2 \log (L_0) = $_____?

Calculate the likelihood ratio chi-square statistic contrasting these two log-likelihood values to asses the likelihood that $\hat{\lambda}$ and $\lambda_0 = 2$ differ by chance alone:

p-value $= $_____?

2. Use the approximate relationship that

variance$(x) = x^2$ variance$(\log[x])$

to show that

if variance$(x) = \dfrac{x^2}{n}$,   then variance$(y) = \dfrac{y^2}{n}$ when $y = \dfrac{1}{x}$.

Basic data set: Leukemia clinical trial data

A clinical trial to evaluate the efficacy of maintenance chemotherapy for acute myelogenous leukemia (AML) was conducted by Embury et al. at Stanford University. After reaching a state of remission through treatment by chemotherapy, the patients who entered the study were randomized into two groups. The first group received maintenance chemotherapy and the second or control group did not. The objective of the trial was to see if maintenance chemotherapy prolonged the time until relapse, that is, increased the length of remission time ("survival time").

These data are to be used for many of the following problems.

Preliminary data collected during the course of the trial are (in weeks):

Maintained ($n_1 = 11$ and $d_1 = 7$)

     remission times : 9, 13, 13^+, 18, 23, 28^+, 31, 34, 45^+, 48, and 161^+

Nonmaintained ($n_0 = 12$ and $d_0 = 11$)

     remission times : 5, 5, 8, 8, 12, 16^+, 23, 27, 30, 33, 43, and 45

$+$ = censored.

Chapter 4. Product-limit estimation

1. Fill in the following table using the leukemia clinical trial data (maintained group), ignoring the censoring and deleting the tied value (13^+) to creat 10 artifical unique and complete survival times.

i	Interval $t_{i-1}-t_i$	Deaths d_i	At-risk n_i	Probability $\hat{q}_i$	Probability $\hat{p}_i$	Survival $\hat{P}_k$	Std. error* $\sqrt{\hat{V}_k}$
1	0–9	——	——	——	——	——	——
2	9–13	——	——	——	——	——	——
3	13–18	——	——	——	——	——	——
4	18–23	——	——	——	——	——	——
5	23–28	——	——	——	——	——	——
6	28–31	——	——	——	——	——	——
7	31–34	——	——	——	——	——	——
8	34–45	——	——	——	——	——	——
9	45–48	——	——	——	——	——	——
10	48–161	——	——	——	——	——	——

$* = \hat{V}_k = \hat{P}_k(1 - \hat{P}_k)/n_k.$

i. Estimate the mean survival time two ways:

     $\hat{\mu}_M = $ _____ ? $\bar{t}_M = $ _____ ?

ii. Estimate the median survival time:

$$\hat{t}_{0.5} = \underline{\hspace{1.5cm}}?$$

iii. Show that for unique and complete data, where $i = 1, 2, 3, \ldots, n$, that

$$\sum \hat{P}_{i-1}(t_i - t_{i-1}) = \sum t_{i-1}(\hat{P}_{i-1} - \hat{P}_i) = \frac{1}{n}\sum t_i = \bar{t}.$$

(note: $\hat{P}_0 = 1$ and $\hat{P}_n = 0$.)

2. Fill in the following table using the leukemia clinical trial data accounting for the influence of the censored data.

				Maintained			
i	Interval $t_{i-1}-t_i$	Deaths d_i	At-risk n_i	Probability $\hat{q}_i$	Probability $\hat{p}_i$	Survival $\hat{P}_k$	Std. error* $\sqrt{\hat{v}_k}$
1	0–9	___	___	___	___	___	___
2	9–13	___	___	___	___	___	___
3	13–18	___	___	___	___	___	___
4	18–23	___	___	___	___	___	___
5	23–31	___	___	___	___	___	___
6	31–34	___	___	___	___	___	___
7	34–48	___	___	___	___	___	___

$* = \hat{v}_k = \hat{P}_k^2 \sum \hat{q}_i/(\hat{p}_i n_i)$ (Greenwood's expression for the variance of $\hat{P}_k$ for $i = 1, 2, \ldots, k$).

i. Estimate the mean survival time $\hat{\mu}_M = \underline{\hspace{1.5cm}}?$

ii. Estimate the survival probability $\hat{P}_5 = \underline{\hspace{1.5cm}}?$

iii. Construct an approximate 95% confidence interval based on the estimated probability $\hat{P}_5$:

lower bound $= \underline{\hspace{2cm}}?$ and upper bound $= \underline{\hspace{2cm}}?$

iv. Estimate the median survival time $\hat{t}_{0.5} = \underline{\hspace{1.5cm}}?$

v. Construct an approximate 95% confidence interval based on the estimated median $\hat{t}_{0.5}$:

lower bound $= \underline{\hspace{2cm}}?$ and upper bound $= \underline{\hspace{2cm}}?$

vi. Plot the estimated survival distribution.

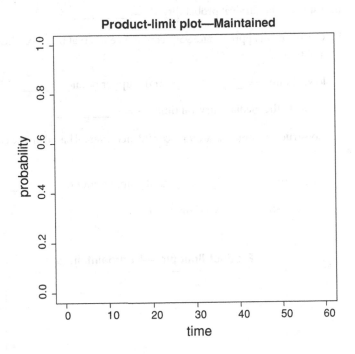

Product-limit plot—Maintained

3. Fill in the following table using the leukemia clinical trial data accounting for the influence of the censored observations.

Nonmaintained

i	Interval $t_{i-1}-t_i$	Deaths d_i	At-risk n_i	Probability $\hat{q}_i$	Probability $\hat{p}_i$	Survival $\hat{P}_k$	Std. error* $\sqrt{\hat{v}_k}$
1	0–5	——	——	——	——	——	——
2	5–8	——	——	——	——	——	——
3	8–12	——	——	——	——	——	——
4	12–23	——	——	——	——	——	——
5	23–27	——	——	——	——	——	——
6	27–30	——	——	——	——	——	——
7	30–33	——	——	——	——	——	——
8	33–43	——	——	——	——	——	——
9	43–45	——	——	——	——	——	——

* See note to previous table.

i. Estimate the mean survival time $\hat{\mu}_{NM}$ = _____ ?

ii. Estimate the survival probability $\hat{P}_5$ = _____ ?

iii. Construct an approximate 95% confidence interval based on the estimated probability $\hat{P}_5$:

lower bound = _____   and   upper bound = _____ ?

iv. Estimate the median survival time $\hat{t}_{0.5}$ = _____ ?

v. Construct an approximate 95% confidence interval based on the estimated median $\hat{t}_{0.5}$:

lower bound = _____   and   upper bound = _____ ?

vi. Plot the estimated survival distribution:

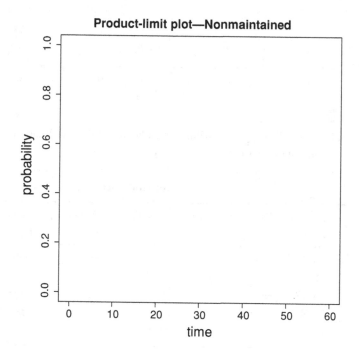

4. **Constant probability of death—q**

Supose $t_{i+1} - t_i = 10$ months and $q_i = q = 0.3$ for all survival times intervals.

i. Fill in the following table:

$t_{i+1} - t_i$	q_i	p_i	P_i
0–10	——	——	——
10–20	——	——	——
20–30	——	——	——
30–40	——	——	——
40–50	——	——	——
50–60	——	——	——
60–70	——	——	——
70–80	——	——	——
80–90	——	——	——
90–100	——	——	——
100–110	——	——	——
120–130	——	——	——

ii. Plot the survival curve.

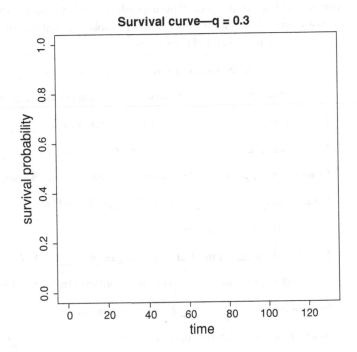

iii. Estimate the mean survival time:

$$\hat{\mu} = \underline{\qquad}?$$

iv. Estimate the median survival time:

$\hat{t}_m = $ _____?

Suppose $t_{i+1} - t_i = \delta$ and $q_i = q$ for all intervals.

v. Create an expression for the survival probability function for the kth interval.

vi. Find a general expression for the mean survival time $\hat{\mu}$.

vii. What is the value of $\hat{\mu}$ when the number of intervals in the table is extremely large?

viii. Estimate $\hat{\mu}$ for $p = 0.7$ when the number of intervals is infinite (time is continuous).

Chapter 5. Exponential survival time probability distribution

1. Using the AML clinical trial data, compare the maintained and nonmaintained treatment of the leukemia patients, assuming each sample is a random sample from exponential survival distribution (constant but possibly different hazard rates).

i. Estimate the mean survival time in both groups:

Maintained: estimated mean survival time $= \hat{\mu}_M = $ _____?

Construct an approximate 95% confidence interval from the estimate:

lower bound $= $ _____? and upper bound $= $ _____?

Nonmaintained: estimated mean survival time $= \hat{\mu}_{NM} = $ _____?

Construct an approximate 95% confidence interval from the estimate:

lower bound $= $ _____? and upper bound $= $ _____?

ii. Estimate the rate of relapse in both groups:

Maintained: estimated survival rate $= \hat{\lambda}_M = $ _____?

Construct an approximate 95% confidence interval from the estimate:

lower bound $= $ _____? and upper bound $= $ _____?

Nonmaintained: estimated survival rate $= \hat{\lambda}_{NM} = $ _____?

Construct an approximate 95% confidence interval from the estimate:

lower bound $= $ _____? and upper bound $= $ _____?

iii. Estimate the median survival time in both groups:

Maintained: estimated median survival time $= \hat{t}_{0.5} =$ _____?

Construct an approximate 95% confidence interval from the estimate:

lower bound = _____? and upper bound = _____?

Nonmaintained: estimated median survival time $= \hat{t}_{0.5} =$ _____?

Construct an approximate 95% confidence interval from the estimate:

lower bound = _____? and upper bound = _____?

iv. Estimate the exponential survival function for both groups and plot these functions $\hat{S}_M(t)$ and $\hat{S}_{NM}(t)$ on the same set of axes.

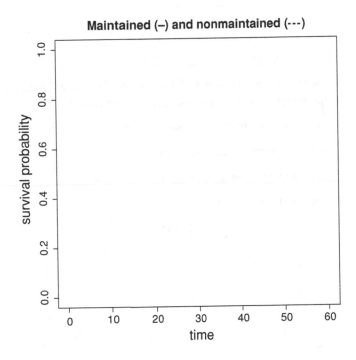

Maintained (–) and nonmaintained (---)

Chapter 6. Weibull survival time probability distribution

1. Using maximum likelihood estimation and a computer algorithm, the estimated scale and shape parameters from the Weibull distribution (λ and γ) applied to the leukemia clinical trial data are as follows:

Maintained:

$\hat{\lambda}_M = 0.0145$ with a standard error of $S_{\hat{\lambda}_M} = 0.0053$

$\hat{\gamma}_M = 1.032$ with a standard error of $S_{\hat{\gamma}_M} = 0.2859$

Nonmaintained:

$\hat{\lambda}_{NM} = 0.0063$ with a standard error of $S_{\hat{\lambda}_{NM}} = 0.0012$.

$\hat{\gamma}_{NM} = 1.574$ with a standard error of $S_{\hat{\gamma}_{NM}} = 0.3841$

i. Construct 95% confidence intervals based on these estimates:

	95% confidence bounds	
	Lower Bound	Upper Bound
$\hat{\lambda}_M$	_____	_____
$\hat{\lambda}_{MN}$	_____	_____
$\hat{\gamma}_M$	_____	_____
$\hat{\gamma}_{MN}$	_____	_____

ii. Estimate the mean and median values based on the estimates $\hat{\lambda}$ and $\hat{\gamma}$ (hint: read the gamma values of $\Gamma(x)$ from the following plot):

maintained: estimated mean value $\hat{\mu}_M =$ _____?

maintained: estimated median value $\hat{t}_{0.5} =$ _____?

nonmaintained: estimated mean value $\hat{\mu}_{NM} =$ _____?

nonmaintained: estimated median value $\hat{t}_{0.5} =$ _____?

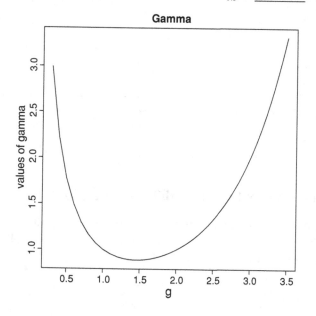

iii. Estimate the hazard functions for the maintained and nonmaintained samples of survival times and plot these two hazard functions on the same set of axes:

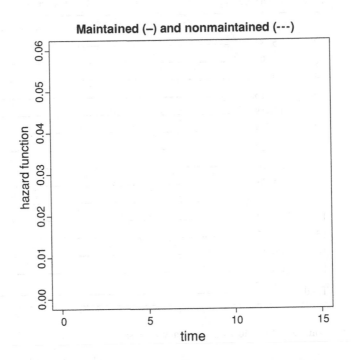

Add to the plot the two hazard functions estimated for the exponential probability distribution (last section—problem set 4).

iv. Is the fit of the Weibull survival distribution to the leukemia data substantially better than the exponential survival distribution? Cite the statistical evidence (test, confidence interval, plots, . . .) to justify your answer.

Chapter 7. Analysis of two-sample survival data

1. Using the AML clinical trial data, perform a log-rank test to assess the difference between maintained and nonmaintained treatment groups.

Log-rank test:
i. Fill in the following table:

i	Interval	a_i	A_i	0 variance(a_i)
1	0–5	————	————	————
2	5–8	————	————	————
3	8–9	————	————	————
4	9–12	————	————	————
5	12–13	————	————	————
6	13–18	————	————	————
7	18–23	————	————	————
8	23–27	————	————	————
9	27–30	————	————	————
10	30–31	————	————	————
11	31–33	————	————	————
12	33–34	————	————	————
13	34–43	————	————	————
14	43–45	————	————	————
15	45–48	————	————	————
Total	—	————	————	————

ii. Compute the following summary values:

$\sum a_i =$ _____? $\quad \sum A_i =$ _____? and variance($\sum a_i$) = _____?
then $X^2 =$ _____? and p-value = _____?

iii. Calculate the values $\hat{c}_i$ and C_i for the 15 intervals (tables) and compare observed sums $\sum a_i$ and $\sum c_i$ to the corresonding expected sums $\sum A_i$ and $\sum C_i$ with chi-square statistic:

then $X^2 =$ _____? and p-value = _____?

2. The following parameter estimates summarize the Weibull hazards survival model applied to the AML clinical trial data:

	Parameter	Estimate	Std. error
Intercept	b_0	−3.180	0.241
Group	b_1	−0.929	0.383
Shape	γ	1.265	0.225

i. Estimate the three selected percentiles for the maintained and nonmaintained groups.

Percentiles	0.10	0.50	0.90
Maintained	————	————	————
Nonmaintained	————	————	————

ii. Calculate the p-value for the comparison of the two groups (maintained versus nonmaintained—$b_1 = 0$?):

$z =$ _____ p-value = _____?

iii. Estimate λ_M and λ_{NM} from the model coefficients $\hat{b}_0$ and $\hat{b}_1$. Demonstrate numerically that $\log(\hat{\lambda}_{NM}/\hat{\lambda}_M) = \hat{b}_1\hat{\gamma}$.

3. i. Plot the estimated survival functions for the maintained and nonmaintained samples of clinical trial data using the product-limit estimates (nonparametric) and the Weibull based estimates (parametric). Plot all four curves on the same set of axes.

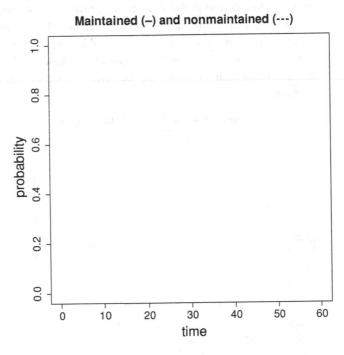

Maintained (–) and nonmaintained (---)

ii. Evaluate the difference in the shape parameters: $\hat{\gamma}_M = 1.032$ with standard error $= 0.286$ and $\hat{\gamma}_{NM} = 1.574$ with standard error $= 0.384$. Note that variance$(x - y)$ = variance(x) + variance(y) and

$$z = \frac{(x - y) - 0}{\sqrt{\text{variance}(x - y)}}$$

has an approximate standard normal distribution when no difference exists between the two estimated value x and y. Therefore, evaluate the hypothesis that $\gamma_M = \gamma_{NM}$:

$X^2 =$ _____? p-value _____?

4. Another approach involves three log-likelihood statistics:

from the maintained group: $-2 \times$ log-likelihood $(\gamma_M) = 71.408$
from the nonmaintained group: $-2 \times$ log-likelihood $(\gamma_{NM}) = 88.288$
from the combined groups: $-2 \times$ log-likelihood $(\gamma) = 161.043$

i. Use these three log-likelihood values to test the equality of shape parameters $(\gamma_M = \gamma_{NM}?)$ and compare the results to the previous test-statistic z^2:

$X^2 =$ _____? p-value _____?

ii. Transform the survival probabilities from the two previously estimated product-limit survival curves (maintained and nonmaintained) and plot the values. That is, plot the pairs of values log$(-\log[\hat{P}_i])$ and log (t_i) for each survival function. Estimate two survival functions $[\hat{S}_{MN}(t)$ and $\hat{S}_N(t)]$ based on the estimated Weibull two-sample model and plot the "log-log" transformed straight lines on the same set of axes.

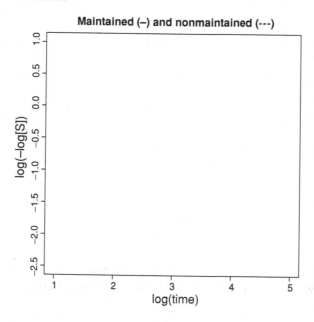

Maintained (–) and nonmaintained (---)

Chapter 8. General hazards model: parametric

1. The leukemia data supplemented with the hypothetical ages of the children partici-
pating in the trial are as follows:
Preliminary data collected during the course of the trial are as follows:

Maintained ($n_1 = 11$)

remission times (in weeks): 9, 13, 13^+, 18, 23, 28^+, 31, 34, 45^+, 48, and 161^+
age (years): 2, 2, 5, 1, 6, 4, 3, 1, 2, 1, and 2

Nonmaintained ($n_1 = 12$)

remission times (in weeks): 5, 5, 8, 8, 12, 16^+, 23, 27, 30, 33, 43, and 45
age (years): 3, 5, 4, 2, 2, 1, 3, 3, 2, 5, 3, and 3

Three hazards models for the leukemia data ($F = \{0 \text{ or } 1\}$ indicates group membership and age $=$ age as reported) are

1. $h(t) = e^{[b_0 + b_1 F + b_2(age - \bar{a}) + b_3 F \times (age - \bar{a})]\gamma} \gamma t^{\gamma - 1}$

2. $h(t) = e^{[b_0 + b_1 F + b_2(age - \bar{a})]\gamma} \gamma t^{\gamma - 1}$

3. $h(t) = e^{[b_0 + b_1 F]\gamma} \gamma t^{\gamma - 1}$

Results from three models are described in the following table:

	Model 1	Model 2	Model 3
Intercept ($\hat{b}_0$)	−3.535	−3.4458	−3.180
Group ($\hat{b}_1$)	0.696	0.858	0.929
Age ($\hat{b}_2$)	−0.114	−0.164	−
Group × age ($\hat{b}_3$)	−0.060	−	−
$\hat{\gamma}$	1.267	1.310	1.264
Log (L)	−80.376	−80.391	−80.521

Note: $\bar{a} = $ mean age of all 23 study participants $= 2.826$ years

i. Comparing log-likelihood values, evaluate the influence of the interaction term in
the model:

$X^2 = $ _____ ? p-value _____ ?

ii. Comparing log-likelihood values, evaluate the influence of the age in the additive
model:

$X^2 = $ _____ ? p-value _____ ?

iii. Comparing log-likelihood values, evaluate the influence of the treatment ignoring
age [hint: $L_0 = -83.179$ when $h(t) = h_0(t)$]:

$X^2 = $ _____ ? p-value _____ ?

iv. Using the additive model including both age and treatment, estimate and plot the baseline survival functions [that is, $F = 0$ = nonmaintained and $age = \bar{a} =$ mean(age)]. In addition, estimate the survival function for the maintained group [that is, $F = 1$ and $age = \bar{a} =$ mean(age)] and plot it on the same set of axes.

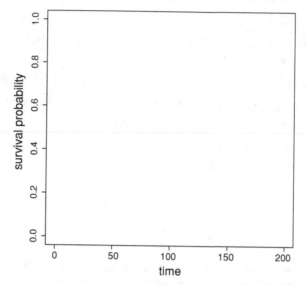

v. Using the additive model including both age and treatment, again compute and plot the baseline survival function [that is, $F = 0$ = nonmaintained and $age = \bar{a} =$ mean(age)]. In addition, estimate the survival function for the nonmaintained group for ages one and five years and plot these curves on the same set of axes (that is, $F = 0$ and $age = 1$ and age $= 5$, where $\bar{a} = 2.826$ years).

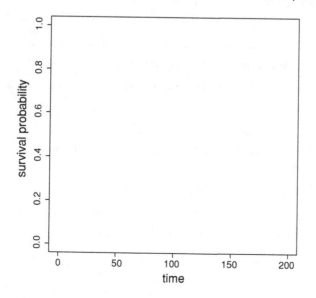

Chapter 9. General hazards model: nonparametric

1. The following again uses the leukemia data supplemented with the hypothetical ages (previous problem).

   Model 1: Two-sample model—$F = 0 =$ nonmaintained and $F = 1 =$ maintained

   $h_1(t) = h_0(t)e^B$

   i. Fill in the table:

	$\hat{B}$	$\hat{S}_{\hat{B}}$	z-score	p-value
Treatments	−0.904	0.512	———	———

   ii. Calculate the approximate 95% confidence for the regression coefficient B (treatments);

   lower bound_____? upper bound_____?

   iii. Estimate the hazard ratio $= hr =$_____?

   iv. Calculate the approximate 95% confidence for the hazard ratio hr:

   lower bound_____? upper bound_____?

   Model 2: Two-variable additive model

   $h_i(t \mid F, \text{age}) = h_0(t)e^{b_1 F + b_2 \text{age}_i}$

   v. Fill in the table:

	$\hat{b}_i$	$\hat{S}_{\hat{b}_i}$	z-score	p-value
Treatments	−0.823	0.531	———	———
Age	0.105	0.203	———	———

   vi. Calculate the approximate 95% confidence for the regression coefficient b_1 (treatments):

   lower bound_____? upper bound_____?

   vii. Estimate the hazard ratio associated with the treatments $= hr =$_____?

   viii. Calculate the approximate 95% confidence for the hazard ratio hr:

   lower bound_____? upper bound_____?

   ix. Calculate the approximate 95% confidence for the regression coefficient b_2 (age):

   lower bound_____? upper bound_____?

x. Calculate the hazard ratio associated with the combined influences of treatment and age $= 5$ years old $(\widehat{HR})$:

$\widehat{HR}$_____?

xi. Calculate the confounding bias associated with the variable age:

bias_____?

xii. Estimate the hazard ratio based on the Weibull additive model applied to the two-sample leukemia data. Compare this hazard ratio with the hazard ratio for the Cox proportional hazards model (treatment only).

xiii. Using these estimates, calculate the hazard ratio $= hr =$_____?

xiv. Compare the hazard ratios from the Cox and Weibull approaches.

2. Survival Curves. Use the nonmaintained group as the "baseline function" and product-limit process to estimate the survival curve $\hat{S}_0(t)$.

i. Estimate the survival curve for the maintained groups $\hat{S}_1(t)$ based on the estimate $\hat{S}_0(t)$ and the two-sample Cox proportional hazards model. Estimate the survival curves for the nonmaintained $\hat{s}_0(t)$ and maintained $\hat{s}_1(t)$ assuming an underlying Weibull distribution (i.e., the parameters previously estimated). Fill in the following table:

Time	Cox model $\hat{S}_0(t)$	$\hat{S}_1(t)$	Weibull model $\hat{s}_0(t)$	$\hat{s}_1(t)$
5	___	___	___	___
8	___	___	___	___
12	___	___	___	___
16	___	___	___	___
23	___	___	___	___
27	___	___	___	___
30	___	___	___	___
33	___	___	___	___
43	___	___	___	___
45	___	___	___	___

ii. Plot both Cox estimated curves [step functions—$\hat{S}_0(t)$ and $\hat{S}_1(t)$] on the same set of axes. Also, plot continuous (smooth line) versions of the Weibull estimated survival curves $\hat{s}_0(t)$ and $\hat{s}_1(t)$ on the same axes.

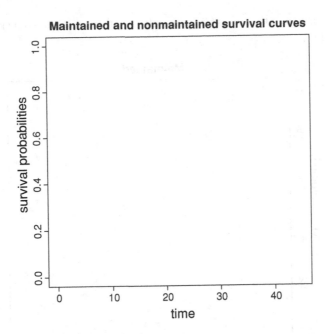

iii. An alternative comparison is achieved by plotting one survival curve against the other. If they are parallel, the plotted points will randomly deviate from a straight line with slope $= 1$ and intercept $= 0$. Plot the points from the table—$\hat{S}_0(t)$ (x-axis) and $\hat{s}_0(t)$ (y-axis). Add a 45° line.

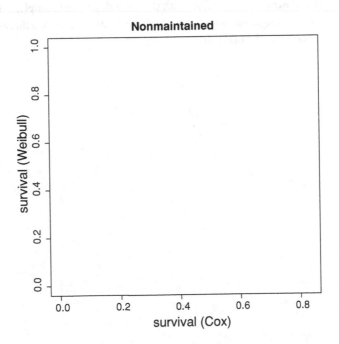

Then, plot the points from the table—$\hat{S}_1(t)$ (x-axis) and $\hat{s}_1(t)$ (y-axis). Add a 45°
line.

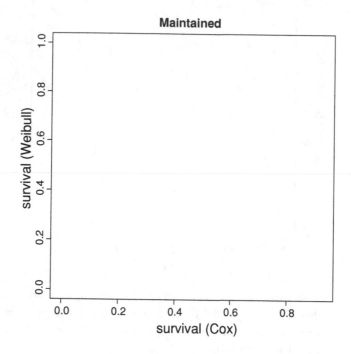

Demonstrate that two lines with the same slope (for example, $a + bx$ and $A + bx$)
form a straight line with slope $= 1.0$ when values from one line are plotted against
values from the other line.

References

References refer to textbooks that give a general description, which is likely more valuable for readers focused on applications. Readers who wish to pursue a topic in more detail can find the original papers in these cited texts.

1. Elant-Johnson, R. C., and Johnson, N. L., *Survival Models and Data Analysis*, John Wiley & Sons, New York, 1980.
2. Hosmer, D. W., and Lemeshow, S., *Applied Survival Analysis*, John Wiley & Sons, New York, 1999.
3. Collett, D., *Modelling Survival Data in Medical Research*, Chapman & Hall, New York, 1994.
4. Chiang, C. L., *The Life Table and Its Applications*, Kreiger Publishing, Malabar, FL, 1984.
5. Miller, R. G., *Survival Analysis*, John Wiley & Sons, New York, 1981.
6. Harrell, F. E., *Regression Modeling Strategies*, Springer-Verlag, New York, 2001.
7. Gross, A. J., and Clark, V. A., *Survival Distributions: Reliability Applications in the Biomedical Sciences*, John Wiley & Sons, New York, 1975.
8. Kalbfleish, J. D., and Prentice, R. L., *The Statistical Analysis of Failure Time Data*, John Wiley & Sons, New York, 1980.
9. Smith, P. J., *Analysis of Failure and Survival Data*, Chapman & Hall, New York, 2002.
10. Tableman, M., and Kim, J. S., *Survival Analysis Using S*, Chapman & Hall, New York, 2004.
11. Cox, D. R., and Oakes, D., *Analysis of Survival Data*, Chapman & Hall, New York, 1985.
12. Lee, E. T., and Wang, J. W., *Survival Methods for Survival Data Analysis*, John Wiley & Sons, New York, 1975.

Index